1989

NONLINEAR METHODS
IN NUMERICAL ANALYSIS

NORTH-HOLLAND MATHEMATICS STUDIES   136
Studies in Computational Mathematics (1)

Editors:

C. Brezinski
University of Lille
Villeneuve d'Ascq, France

L. Wuytack
University of Antwerp
Wilrijk, Belgium

NORTH-HOLLAND – AMSTERDAM ● NEW YORK ● OXFORD ● TOKYO

NONLINEAR METHODS IN NUMERICAL ANALYSIS

Annie CUYT
Luc WUYTACK
University of Antwerp
Belgium

NORTH-HOLLAND – AMSTERDAM ● NEW YORK ● OXFORD ● TOKYO

ISBN: 0 444 70189 3

First edition 1987
Second printing 1988

Publishers:
ELSEVIER SCIENCE PUBLISHERS B.V.
P.O. BOX 1991
1000 BZ AMSTERDAM
THE NETHERLANDS

Sole distributors for the U.S.A. and Canada:
ELSEVIER SCIENCE PUBLISHING COMPANY, INC.
52 VANDERBILT AVENUE
NEW YORK, N.Y. 10017
U.S.A.

Library of Congress Cataloging-in-Publication Data

Cuyt, Annie, 1956-
   Nonlinear methods in numerical analysis.

   (North-Holland mathematics studies ; 136) (Studies in
computational mathematics ; 1)
   Bibliography: p.
   Includes index.
   1. Numerical analysis. I. Wuytack, L. (Luc), 1943-
II. Title. III. Series. IV. Series: Studies in
computational mathematics ; 1.
QA297.C89 1987   519.4   86-32932
ISBN O-444-70189-3 (U.S.)

PRINTED IN THE NETHERLANDS

To Annelies Van Soom from her mother.
Annie Cuyt

PREFACE.

Most textbooks on Numerical Analysis discuss linear techniques for the solution of various numerical problems. Only a small number of books introduce and illustrate nonlinear methods.

This book accumulates several nonlinear techniques mainly resulting from the use of Padé approximants and rational interpolants. First these types of rational approximants are introduced and afterwards methods based on their use are developed for the solution of standard problems in numerical mathematics : convergence acceleration, initial value problems, boundary value problems, quadrature, nonlinear equations, partial differential equations and integral equations. The problems are allowed to be univariate or multivariate.

The treatment of the univariate theory results from a course given by the second author at the University of Leuven and completed by the first author with many new theorems and numerical results. The discussion of the multivariate theory is based on research work by the first author. The text as it stands is now used for a graduate course in Numerical Analysis at the University of Antwerp. The book brings together many results of research work carried out at the University of Antwerp during the past few years. We particularly mention results of Guido Claessens, Albert Wambecq, Paul Van der Cruyssen and Brigitte Verdonk.

Let us now give a survey of the contents of this book and a motivation for the problems treated.

Since continued fractions play an important role, **Chapter I** is an introduction to this topic. We mention some basic properties, evaluation algorithms and convergence theorems. From the section dealing with convergence we can already learn that in certain situations nonlinear approximations are more powerful than linear approximations. The recent notion of branched continued fraction is introduced in the multivariate section and will be used for the construction of multivariate rational interpolants.

In **Chapter II** Padé approximants are treated. They are local rational approximants for a given function. The problems of existence, unicity and computation are treated in detail. Also the convergence of sequences of Padé approximants and the continuity of the Padé operator which associates with a function its Padé approximant of a certain order, are considered. Again a special section is devoted to the multivariate case. We do not discuss the relationship between Padé approximants and orthogonal polynomials or the moment problem.

In **Chapter III** rational interpolants are defined. Their function values fit those of a given function at some given points. Many results of the previous chapter remain valid for this more general case where the interpolation conditions are spread over several points. In between the rational interpolation case and the Padé approximation case lies the theory of rational Hermite interpolation where each interpolation point can be assigned more than one interpolation condition. Some results on the convergence of sequences of rational Hermite interpolants are mentioned and multivariate rational interpolants are introduced in two different ways.

The previous types of rational approximants are used in **Chapter IV** to develop several numerical methods for the solution of classical problems such as convergence acceleration, nonlinear equations, ordinary differential equations, numerical quadrature, partial differential equations and integral equations. Many numerical examples illustrate the different techniques and we see that the nonlinear methods are very useful in situations where we are faced with singularities. However, one must be careful in applying the nonlinear methods due to the fact that denominators in the formula can get small.

We tried to make the text as self-contained as possible. Each chapter also contains a problem section and a section with remarks that indicate extensions of the discussed theory. References to the literature are given at the end of each chapter in alphabetical order. In the text we refer to them within square brackets. Formulas and equations are numbered as $(a.b.)$, where a indicates the chapter number and b the number of the formula in that chapter.

In preparing the text the authors did benefit from discussions with many colleagues and friends. We mention in particular Claude Brezinski (Lille), Marcel de Bruin (Amsterdam), William Gragg (Lexington), Peter Graves-Morris (Canterbury), Louis Rall (Madison), Nico Temme (Amsterdam), Helmut Werner (Bonn).

We also thank Drs. A. Sevenster from North Holland Publishing Co who encouraged us to write this book and Mrs. F. Schoeters and Mrs. R. Vanmechelen who typed the manuscript.

Antwerp

Annie Cuyt
Luc Wuytack

NONLINEAR METHODS
IN
NUMERICAL ANALYSIS

CHAPTER I: Continued Fractions.

"J'ai eu l'honneur de présenter à l'Académie en 1802 un mémoire sous le titre : Essai d'une méthode générale pour réduire toutes sortes de séries en fractions continues. Après ce temps ayant eu occasion de penser encore à cette matière, j'ai fait de nouvelles réflexions qui peuvent servir à perfectionner et simplifier la méthode dont il s'agit. Ce sont ces réflexions que je présente maintenant à la société savante."

B. VISCOVATOV — *"De la méthode générale pour réduire toutes sortes de quantités en fractions continues"* *(1805).*

§1. Notations and definitions.

A **continued fraction** is an expression of the form

$$b_0 + \cfrac{a_1}{b_1 + \cfrac{a_2}{b_2 + \cfrac{a_3}{b_3 + \ldots}}}$$

$$\ldots + \cfrac{a_i}{b_i + \ldots}$$

where the a_i and b_i are real (or complex) numbers or functions and are respectively called **partial numerators** and **partial denominators**. Instead of the expression above we will most of the times use the following compact notations:

$$b_0 + \frac{a_1|}{|b_1} + \frac{a_2|}{|b_2} + \frac{a_3|}{|b_3} + \ldots + \frac{a_i|}{|b_i} + \ldots$$

or

$$b_0 + \sum_{i=1}^{\infty} \frac{a_i|}{|b_i} \qquad\qquad (1.1.)$$

The truncation

$$C_n = b_0 + \sum_{i=1}^{n} \frac{a_i|}{|b_i} \qquad\qquad n = 0, 1, 2, \ldots$$

is called the nth **convergent** of the continued fraction (1.1.). If

$$\lim_{n \to \infty} \ C_n = C$$

exists and is finite, then the continued fraction is said to be **convergent** and C is called the **value** of the continued fraction. Clearly C_n is a rational expression

$$C_n = \frac{P_n}{Q_n} = \frac{P_n(b_0, a_1, b_1, \ldots, a_n, b_n)}{Q_n(b_0, a_1, b_1, \ldots, a_n, b_n)} \qquad\qquad (1.2.)$$

where P_n and Q_n are polynomials of a certain degree in the $2n + 1$ partial numerators and denominators $b_0, a_1, b_1, \ldots, a_n, b_n$. The polynomials P_n and Q_n are respectively called the nth **numerator** and n^{th} **denominator** of the continued fraction (1.1.).

§2. Fundamental properties.

2.1. Recurrence relations for P_n and Q_n.

The nth numerators and denominators satisfy the same three-term recurrence relation, but with different starting values. This relation is given in the following theorem.

Theorem 1.1.

If $P_{-1} = 1$, $P_0 = b_0$, $Q_{-1} = 0$, $Q_0 = 1$, then for $n \geq 1$

$$\begin{cases} P_n = b_n \ P_{n-1} + a_n \ P_{n-2} \\ Q_n = b_n \ Q_{n-1} + a_n \ Q_{n-2} \end{cases} \tag{1.3.}$$

Proof

The proof is performed by induction.
Obviously

$$C_1 = \frac{b_1 . b_0 + a_1 . 1}{b_1 . 1 + a_1 . 0}$$

and so the formulas (1.3.) are valid for $n = 1$. Let us now suppose the validity of (1.3.) for $n \leq k$. We will prove it for $n = k + 1$.
We have, using (1.2.),

$$\begin{aligned} C_{k+1} &= \frac{P_{k+1}(b_0, a_1, b_1, \ldots, a_k, b_k, a_{k+1}, b_{k+1})}{Q_{k+1}(b_0, a_1, b_1, \ldots, a_k, b_k, a_{k+1}, b_{k+1})} \\ &= \frac{P_k(b_0, a_1, b_1, \ldots, a_k, b_k + \dfrac{a_{k+1}}{b_{k+1}})}{Q_k(b_0, a_1, b_1, \ldots, a_k, b_k + \dfrac{a_{k+1}}{b_{k+1}})} \end{aligned}$$

Consequently, by using (1.3.) for $n = k$,

$$C_{k+1} = \frac{\left(b_k + \dfrac{a_{k+1}}{b_{k+1}}\right) P_{k-1} + a_k\ P_{k-2}}{\left(b_k + \dfrac{a_{k+1}}{b_{k+1}}\right) Q_{k-1} + a_k\ Q_{k-2}}$$

$$= \frac{b_{k+1}(b_k\ P_{k-1} + a_k\ P_{k-2}) + a_{k+1}\ P_{k-1}}{b_{k+1}(b_k\ Q_{k-1} + a_k\ Q_{k-2}) + a_{k+1}\ Q_{k-1}}$$

$$= \frac{b_{k+1}\ P_k + a_{k+1}\ P_{k-1}}{b_{k+1}\ Q_k + a_{k+1}\ Q_{k-1}}$$

■

2.2. Euler-Minding series.

It is easy now to give an expression for the difference of two consecutive convergents of a continued fraction.

Theorem 1.2.

If $Q_n\ Q_{n-1} \neq 0$, then

$$C_n - C_{n-1} = (-1)^{n+1} \frac{a_1\ a_2 \ldots a_n}{Q_n\ Q_{n-1}} \tag{1.4.}$$

Proof

One can show, by induction and using the recurrence relations (1.3.), that for $n \geq 1$

$$P_n\ Q_{n-1} - Q_n\ P_{n-1} = (-1)^{n+1} a_1\ a_2 \ldots a_n$$

From this (1.4.) follows immediately. ■

This theorem can now be used to give a more explicit formula for C_n.

Theorem 1.3.

If $Q_i \neq 0$ for $1 \leq i \leq n$, then

$$C_n = b_0 + \sum_{i=1}^{n} (-1)^{i+1} \frac{a_1 \ldots a_i}{Q_{i-1} Q_i} \tag{1.5.}$$

Proof

We have, by means of formula (1.4.),

$$\frac{P_n}{Q_n} = \left(\frac{P_n}{Q_n} - \frac{P_{n-1}}{Q_{n-1}} \right) + \left(\frac{P_{n-1}}{Q_{n-1}} - \frac{P_{n-2}}{Q_{n-2}} \right) + \ldots + \left(\frac{P_1}{Q_1} - \frac{P_0}{Q_0} \right) + \frac{P_0}{Q_0}$$

$$= (-1)^{n+1} \frac{a_1 \ldots a_n}{Q_n Q_{n-1}} + (-1)^n \frac{a_1 \ldots a_{n-1}}{Q_{n-1} Q_{n-2}} + \ldots + \frac{a_1}{Q_1 Q_0} + b_0$$

∎

The expression (1.5.) is the nth partial sum of the series

$$b_0 + \sum_{i=1}^{\infty} (-1)^{i+1} \frac{a_1 \ldots a_i}{Q_{i-1} Q_i}$$

which is called the **Euler-Minding series.**
Thus we have associated a series with a continued fraction such that the nth partial sum of the series equals the nth convergent of the continued fraction. This interrelation between series and continued fractions can be used to apply well-known results for series to the theory of continued fractions.

2.3. Equivalence transformations.

Let $p_i \neq 0$ for $i \geq 0$. The transformation that alters the continued fraction (1.1.) into

$$b_0 + \frac{p_1 a_1 \vert}{\vert p_1 b_1} + \sum_{i=2}^{\infty} \frac{p_{i-1} p_i a_i \vert}{\vert p_i b_i} \tag{1.6.}$$

is called an **equivalence transformation.** Clearly (1.1.) and (1.6.) have the same convergents. By performing equivalence transformations a continued fraction can be rewritten in a prescribed form. For instance, if $a_i \neq 0$ for $i \geq 1$, then (1.1.) can be rewritten as

$$d_0 + \sum_{i=1}^{\infty} \frac{1}{\left| d_i \right.} \tag{1.7.}$$

by choosing

$$p_1 = \frac{1}{a_1}$$

and

$$p_i = \frac{1}{a_i \, p_{i-1}} \quad \text{for} \quad i \geq 2.$$

Hence one can limit the study of continued fractions to continued fractions of the form (1.7.). Such a continued fraction is called a **reduced continued fraction** [10 p. 480].

2.4. Contraction of a continued fraction.

Let us consider the following problem. Suppose we are given a sequence $\{C_n\}_{n \in \mathbb{N}}$ of subsequently different elements and we want to construct a continued fraction of which C_n is the nth convergent.

Theorem 1.4.

If $C_n \neq C_{n-1}$ for $n \geq 1$, then the continued fraction

$$b_0 + \sum_{i=1}^{\infty} \frac{a_i}{\left| b_i \right.}$$

with

$$b_0 = C_0$$
$$a_1 = C_1 - C_0$$
$$b_1 = 1$$

and with for $i \geq 2$

$$a_i = \frac{C_{i-1} - C_i}{C_{i-1} - C_{i-2}}$$

$$b_i = \frac{C_i - C_{i-2}}{C_{i-1} - C_{i-2}}$$

has the elements of the sequence $\{C_n\}_{n \in \mathbb{N}}$ as convergents.

Proof

We write

$$C_n = \frac{P_n}{Q_n}$$

with $P_n = C_n$ and $Q_n = 1$.

Since the partial numerators and denominators of the continued fraction

$$b_0 + \sum_{i=1}^{\infty} \frac{a_i}{\vert b_i}$$

with convergents C_n must satisfy the relations (1.3.), we get the following system of equations in the unknowns b_0, a_i and b_i:

$$b_0 = C_0$$
$$b_1\, b_0 + a_1 = C_1$$
$$b_1 = 1$$
$$\text{for } i \geq 2 : \begin{cases} b_i\, C_{i-1} + a_i\, C_{i-2} = C_i \\ b_i + a_i = 1 \end{cases}$$

A solution of this system is given by

$$b_0 = C_0$$
$$a_1 = C_1 - C_0$$
$$b_1 = 1$$
$$\text{for } i \geq 2 : \begin{cases} a_i = \dfrac{C_{i-1} - C_i}{C_{i-1} - C_{i-2}} \\[2mm] b_i = \dfrac{C_i - C_{i-2}}{C_{i-1} - C_{i-2}} \end{cases} \qquad (1.8.)$$

∎

If also $C_n \neq C_{n-2}$ for $n \geq 2$, then by means of an equivalence transformation the continued fraction with partial numerators and denominators given by (1.8.), can be written as

$$C_0 + \frac{C_1 - C_0}{\vert 1} + \frac{C_1 - C_2}{\vert C_2 - C_0} + \sum_{i=3}^{\infty} \frac{(C_{i-2} - C_{i-3})(C_{i-1} - C_i)}{\vert C_i - C_{i-2}} \qquad (1.9.)$$

The formulas (1.8.) and (1.9.) can be used to compute a **contraction** of a continued fraction, i.e. a continued fraction constructed in such a way that its convergents form a subsequence of the sequence of convergents of the given continued fraction. We shall now illustrate this.

2.5. Even and odd part.

Consider a continued fraction with convergents $\{C_n\}_{n \in \mathbb{N}}$.
The **even part** of this continued fraction is a continued fraction with convergents $\{C_{2n}\}_{n \in \mathbb{N}}$, while the **odd part** is a continued fraction with convergents $\{C_{2n+1}\}_{n \in \mathbb{N}}$. Theorem 1.4. enables us to construct those even and odd parts. We now derive a formula for the even part and give an analogous formula for the odd part without proof.
Consider the continued fraction

$$b_0 + \sum_{i=1}^{\infty} \frac{a_i|}{|b_i}$$

with convergents

$$\{\frac{P_i}{Q_i}\}_{i \in \mathbb{N}}$$

The partial numerators and denominators of the even part as expressed in the partial numerators and denominators a_i and b_i, are computed as follows.
Let

$$C_i = \frac{P_{2i}}{Q_{2i}}$$

use the formulas (1.8.) and perform an equivalence transformation with

$$p_i = \frac{Q_{2i}}{Q_{2(i-1)}}$$

We then get the following continued fraction

$$\frac{P_0}{Q_0} + \frac{\left(\dfrac{P_2}{Q_2} - \dfrac{P_0}{Q_0}\right)\dfrac{Q_2}{Q_0}\bigg|}{\left|\dfrac{Q_2}{Q_0}\right.} + \sum_{i=2}^{\infty} \frac{d_i|}{|e_i}$$

with

$$d_i = \frac{Q_{2i}}{Q_{2(i-2)}} \left(\frac{P_{2(i-1)}}{Q_{2(i-1)}} - \frac{P_{2i}}{Q_{2i}} \right) \bigg/ \left(\frac{P_{2(i-1)}}{Q_{2(i-1)}} - \frac{P_{2(i-2)}}{Q_{2(i-2)}} \right)$$

and

$$e_i = \frac{Q_{2i}}{Q_{2(i-1)}} \left(\frac{P_{2i}}{Q_{2i}} - \frac{P_{2(i-2)}}{Q_{2(i-2)}} \right) \bigg/ \left(\frac{P_{2(i-1)}}{Q_{2(i-1)}} - \frac{P_{2(i-2)}}{Q_{2(i-2)}} \right)$$

which is equal to

$$b_0 + \frac{P_2 - b_0\,Q_2}{Q_2} \bigg| + \sum_{i=2}^{\infty} \frac{\dfrac{P_{2(i-1)}\,Q_{2i} - Q_{2(i-1)}\,P_{2i}}{P_{2(i-1)}\,Q_{2(i-2)} - Q_{2(i-1)}\,P_{2(i-2)}}}{\dfrac{P_{2i}\,Q_{2(i-2)} - Q_{2i}\,P_{2(i-2)}}{P_{2(i-1)}\,Q_{2(i-2)} - Q_{2(i-1)}\,P_{2(i-2)}}} \bigg|$$

Analogous to formula (1.4.) one can prove that

$$\frac{P_{2i}}{Q_{2i}} - \frac{P_{2(i-1)}}{Q_{2(i-1)}} = \frac{(-1)^{2i}\,a_1 \ldots a_{2i-1}\,b_{2i}}{Q_{2i}\,Q_{2(i-1)}}$$

and that

$$\frac{P_{2(i+1)}}{Q_{2(i+1)}} - \frac{P_{2(i-1)}}{Q_{2(i-1)}} = \frac{(-1)^{2i}\,a_1 \ldots a_{2i-1} \left[b_{2i}\left(b_{2i+1}\,b_{2i+2} + a_{2i+2} \right) + b_{2i+2}\,a_{2i+1} \right]}{Q_{2(i+1)}\,Q_{2(i-1)}}$$

(see problems (1) and (2) at the end of this chapter).
Finally we get for the even part

$$b_0 + \frac{a_1\,b_2}{b_1\,b_2 + a_2} \bigg| - \frac{a_2\,a_3\,b_4}{(b_2\,b_3 + a_3)\,b_4 + b_2\,a_4} \bigg|$$

$$+ \sum_{i=3}^{\infty} \frac{-a_{2i-2}\,a_{2i-1}\,b_{2i-4}\,b_{2i}}{b_{2i}(b_{2i-2}\,b_{2i-1} + a_{2i-1}) + b_{2i-2}\,a_{2i}} \bigg| \tag{1.10a.}$$

In the same way one can prove that the odd part is

$$
\frac{b_0\, b_1 + a_1}{b_1} - \left.\frac{a_1\, a_2\, b_3/\, b_1}{(b_1\, b_2 + a_2)b_3 + b_1\, a_3}\right|
$$

$$
+ \sum_{i=2}^{\infty} \left.\frac{-a_{2i}\, a_{2i-1}\, b_{2i-3}\, b_{2i+1}}{(b_{2i}b_{2i+1} + a_{2i+1})b_{2i-1} + a_{2i}\, b_{2i+1}}\right| \qquad (1.10\text{b.})
$$

We illustrate this procedure with the following example.
To compute $C = \sqrt{r}$ with r real positive, we first write $r = b^2 + a$ with a positive. So $C^2 - b^2 = a$ or

$$
C = b + \frac{a}{b + C}
$$

Hence

$$
C = b + \sum_{i=1}^{\infty} \left.\frac{a}{2b}\right|
$$

If we take $r = 102$, we get

$$
\sqrt{102} = 10 + \sum_{i=1}^{\infty} \left.\frac{2}{20}\right|
$$

The even part of this continued fraction is

$$
10 + \left.\frac{20}{201}\right| + \sum_{i=1}^{\infty} \left.\frac{-1}{202}\right|
$$

and the odd part is

$$
\frac{101}{10} - \left.\frac{1}{2020}\right| - \left.\frac{10}{202}\right| + \sum_{i=1}^{\infty} \left.\frac{-1}{202}\right|
$$

§3. Methods to construct continued fractions.

We are now going to describe some methods that can be used to write a given number or function as a continued fraction. Other techniques can for example be found in [10 pp. 487-500] and [12 pp. 76-150]. Some convergence problems of such continued fractions are treated in the next section.

3.1. Successive substitution.

Let f be a given number or function. Write $T_0 = f$ and compute $T_1, T_2, \ldots, T_{n+1}$ such that

$$T_0 = b_0 + T_1$$
$$T_1 = \frac{a_1}{b_1 + T_2}$$
$$T_i = \frac{a_i}{b_i + T_{i+1}} \qquad 2 \leq i \leq n$$

where b_0, a_i and b_i are chosen freely and can be functions of the argument of f. In this way we get

$$f = T_0 = b_0 + T_1 = b_0 + \sum_{i=1}^{n-1} \left.\frac{a_i}{b_i}\right| + \left.\frac{a_n}{b_n + T_{n+1}}\right|$$

By continuing this method of successive substitution we get an expression of the form (1.1.). It is important to check whether f is really the value of this continued fraction or for which arguments of f this is true. Such problems are treated further on.

We shall illustrate this method by calculating a continued fraction expansion for e^x. Since

$$e^x = 1 + x + \frac{x^2}{2} + \frac{x^3}{6} + \ldots$$

we choose $b_0 = 1$.

So we get

$$T_1 = e^x - 1 = x\left(1 + \frac{x}{2} + \frac{x^2}{6} + \ldots\right)$$

$$= \frac{x}{1 - \left(\dfrac{x}{2} - \dfrac{x^2}{12} + \ldots\right)}$$

$$= \frac{x}{1 - x + \left(\dfrac{x}{2} + \dfrac{x^2}{12} + \ldots\right)}$$

which suggests us to write

$$T_1 = \frac{a_1}{b_1 + T_2}$$

with

$$a_1 = x$$
$$b_1 = 1 - x$$
$$T_2 = \frac{x}{2}\left(1 + \frac{x}{6} + \ldots\right)$$

Some easy computations show that

$$T_2 = \frac{x}{2}\frac{1}{1 - \left(\dfrac{x}{6} + \ldots\right)}$$

$$= \frac{x}{2 - \left(\dfrac{x}{3} + \ldots\right)}$$

$$= \frac{x}{2 - x + \left(\dfrac{2x}{3} + \ldots\right)}$$

So we can write

$$T_2 = \frac{a_2}{b_2 + T_3}$$

with

$$a_2 = x$$
$$b_2 = 2 - x$$
$$T_3 = \frac{2x}{3}(1 + \ldots)$$

If we continue in this way, we find that for $a_i = (i-1)x$ and $b_i = i - x$,

$$T_{i+1} = \frac{ix}{i+1}\left(1 + \sum_{k=1}^{\infty} e_k\, x^k\right)$$

which again suggests the choice $a_{i+1} = ix$ and $b_{i+1} = (i+1) - x$, such that

$$e^x = 1 + \frac{x}{\big|1-x} + \sum_{i=2}^{n-1} \frac{(i-1)x}{\big|i-x} + \frac{(n-1)x}{\big|n-x+T_{n+1}}$$

or by continuing the process [12 p.130]

$$e^x = 1 + \frac{x}{\big|1-x} + \sum_{i=2}^{\infty} \frac{(i-1)x}{\big|i-x} \tag{1.11.}$$

Another example is given by the construction of a continued fraction for

$$C = \sqrt{f(x,y)} = \sqrt{x^2 + 3xy + y^2}$$

with x and y real positive. Proceeding as in the previous section we can write

$$C^2 = f(x,y) = b^2(x,y) + a(x,y)$$

with $b(x,y) = x + y$ and $a(x,y) = xy$. Then

$$C = b + \frac{a}{b+C}$$

and consequently

$$\sqrt{x^2 + 3xy + y^2} = (x+y) + \sum_{i=1}^{\infty} \frac{xy}{\big|2(x+y)}$$

3.2. Equivalent continued fractions.

A series

$$\sum_{i=0}^{\infty} d_i$$

and a continued fraction

$$b_0 + \sum_{i=1}^{\infty} \left.\frac{a_i}{b_i}\right|$$

are called **equivalent** if for every $n \geq 0$ the nth partial sum

$$D_n = \sum_{i=0}^{n} d_i$$

of the series equals the nth convergent

$$C_n = b_0 + \sum_{i=1}^{n} \left.\frac{a_i}{b_i}\right|$$

Remember that the Euler-Minding series and the continued fraction (1.1.) are equivalent. Transforming a given series into an equivalent continued fraction can also be done by means of formula (1.9.), with

$$C_n = \sum_{i=0}^{n} d_i$$

We obtain

$$d_0 + \left.\frac{d_1}{1}\right| + \left.\frac{-d_2}{d_1 + d_2}\right| + \sum_{i=3}^{\infty} \left.\frac{-d_{i-2}\, d_i}{d_{i-1} + d_i}\right| \qquad (1.12.)$$

In the following example $\sum_{i=0}^{\infty} d_i$ is a power series.
Consider

$$e^x = \sum_{i=0}^{\infty} \frac{x^i}{i!}$$

Then after an equivalence transformation (1.12.) is

$$1 + \left.\frac{x}{1}\right| + \sum_{i=2}^{\infty} \left.\frac{-\dfrac{1}{(i-2)!\,i!}\,x}{\dfrac{1}{(i-1)!} + \dfrac{1}{i!}x}\right|$$

If we perform another equivalence transformation, we get

$$e^x = 1 + \left.\frac{x}{1}\right| + \sum_{i=2}^{\infty} \left.\frac{-(i-1)x}{i+x}\right| \tag{1.13.}$$

Remark that an equivalent continued fraction will converge if and only if the given power series converges. For our example this means that (1.13.) converges on the whole complex plane. Hence, by substitution of x by $-x$,

$$\sum_{i=2}^{\infty} \left.\frac{(i-1)x}{i-x}\right|$$

converges for all x.

Consequently the continued fraction in the righthand side of (1.11.) converges for arbitrary x.

3.3. The method of Viscovatov.

This method is used to develop a continued fraction expansion for functions given as the quotient of two power series [27].

Let

$$f(x) = \frac{d_{10} + d_{11}x + d_{12}x^2 + \dots}{d_{00} + d_{01}x + d_{02}x^2 + \dots}$$

Then

$$f(x) = \cfrac{1}{\dfrac{d_{00}}{d_{10}} + \dfrac{d_{00} + d_{01}x + d_{02}x^2 + \dots}{d_{10} + d_{11}x + d_{12}x^2 + \dots} - \dfrac{d_{00}}{d_{10}}}$$

$$= \cfrac{d_{10}}{d_{00} + x\,\dfrac{(d_{10}d_{01} - d_{00}d_{11}) + (d_{10}d_{02} - d_{00}d_{12})x + \dots}{d_{10} + d_{11}x + d_{12}x^2 + \dots}}$$

If we put $d_{2,i} = d_{10} d_{0,i+1} - d_{00} d_{1,i+1}$ for $i \geq 0$, we get

$$f(x) = \cfrac{d_{10}}{d_{00} + x \cfrac{d_{20} + d_{21}x + d_{22}x^2 + \ldots}{d_{10} + d_{11}x + d_{12}x^2 + \ldots}}$$

This procedure can be repeated and if we define

$$d_{k,i} = d_{k-1,0} d_{k-2,i+1} - d_{k-2,0} d_{k-1,i+1}$$

for $k > 2$ and $i \geq 0$, we finally get

$$f(x) = \left. \frac{d_{10}}{d_{00}} \right| + \left. \frac{d_{20}x}{d_{10}} \right| + \left. \frac{d_{30}x}{d_{20}} \right| + \ldots \tag{1.14.}$$

In case $d_{0i} = 0$ for $i \geq 1$ then $f(x)$ is a power series itself and we shall prove in the next section that the method of Viscovatov can be used to compute a corresponding continued fraction.

If $f(x)$ is the quotient of two polynomials and thus a rational function, this method can be used to write f in the form of a continued fraction (see also problem (8)).

3.4. Corresponding and associated continued fractions.

A continued fraction

$$b_0(x) + \sum_{i=1}^{\infty} \left. \frac{a_i(x)}{b_i(x)} \right|$$

for which the Taylor series development of the nth convergent $C_n(x)$ around the origin matches a given power series

$$\sum_{i=0}^{\infty} c_i x^i$$

up to and including the term of degree n is called **corresponding** to this power series. In other words, for a corresponding continued fraction, if

$$C_n(x) = b_0(x) + \sum_{i=1}^{n} \left. \frac{a_i(x)}{b_i(x)} \right| = \sum_{i=0}^{\infty} e_i x^i$$

then for every n we have $e_i = c_i$ for $i = 0, \ldots, n$. A lot of methods to construct corresponding continued fractions are treated in chapter II.

A continued fraction

$$b_0(z) + \sum_{i=1}^{\infty} \frac{a_i(z)|}{|b_i(z)|}$$

for which the Taylor series development of the nth convergent $C_n(z)$ matches a given power series

$$\sum_{i=0}^{\infty} c_i \, x^i$$

up to and including the term of degree $2n$ is called **associated**. A corresponding continued fraction can be turned into an associated one by calculating the even part.

Let us now again consider the algorithm of Viscovatov. For the continued fraction (1.14.) we define

$$f_0 = f \text{ given by (1.14.)}$$
$$f_1 = d_{10} - d_{00} f_0$$
$$f_k = d_{k,0} x \, f_{k-2} - d_{k-1,0} \, f_{k-1} \qquad k = 2, 3, 4, \ldots$$

Then by induction it is easy to see that $f_k(x)$ can be developed into a series of the form

$$f_k(x) = x^k \sum_{i=0}^{\infty} e_i^{(k)} x^i$$

One can also prove by induction that for the kth convergent

$$\frac{P_k}{Q_k} = \frac{d_{10}|}{|d_{00}} + \frac{d_{20} x|}{|d_{10}} + \frac{d_{30} x|}{|d_{20}} + \ldots + \frac{d_{k,0} x|}{|d_{k-1,0}}$$

the relation

$$f - \frac{P_k}{Q_k} = (-1)^k \, \frac{f_k}{Q_k}$$

holds.

Hence if $f(x)$ is given by the series expansion

$$f(x) = c_0 + c_1 x + c_2 x^2 + \ldots$$

the algorithm of Viscovatov when applied to $(f(x) - c_0)/x$ with $d_{1i} = c_{1+i}$ for $i \geq 0$, generates a continued fraction of the form

$$\frac{c_1}{\vert 1} + \frac{d_{20}\, x}{\vert c_1} + \frac{d_{30}\, x}{\vert d_{20}} + \ldots$$

of which the Taylor series development of the kth convergent matches the power series $(f(x) - c_0)/x$ up to and including the term of degree $k - 1$. In this way we obtain for $f(x)$ the corresponding continued fraction

$$c_0 + \frac{c_1 x}{\vert 1} + \frac{d_{20}\, x}{\vert c_1} + \frac{d_{30}\, x}{\vert d_{20}} + \ldots$$

because the Taylor series development of the kth convergent matches the power series $f(x)$ up to and including the term of degree k.

3.5. Thiele interpolating continued fractions.

This technique is dealt with completely in chapter III. It uses interpolation data and reciprocal differences. Thiele type continued fractions will be constructed both for univariate and multivariate functions. In the univariate case continued fractions of the form

$$b_0 + \sum_{i=1}^{\infty} \frac{a_i(x - x_i)}{\vert b_i}$$

will be used while we shall need branched continued fractions similar to those in (1.29.) for the multivariate case.

§4. Convergence of continued fractions.

4.1. Convergence criteria.

The following result is a classical convergence criterion for reduced continued fractions, and is due to Seidel [21]. It dates from 1846.

Theorem 1.5.

If $b_i > 0$ for $i \geq 1$, then the continued fraction

$$b_0 + \sum_{i=1}^{\infty} \frac{1|}{|b_i}$$

converges, if and only if the series $\sum_{i=1}^{\infty} b_i$ diverges.

Proof

The Euler-Minding series for

$$b_0 + \sum_{i=1}^{\infty} \frac{1|}{|b_i}$$

is

$$b_0 + \sum_{i=1}^{\infty} \frac{(-1)^{i+1}}{Q_{i-1}\, Q_i}$$

It is an alternating series because $a_i = 1$ and $b_i > 0$ imply that $Q_i > 0$ for $i \geq 1$. The nth denominator Q_n is bounded below by $\theta = \min(1, b_1)$. This can be proved by induction from the recurrence relation for Q_n.

If we put $r_n = Q_n\, Q_{n-1}$ then the r_n are monotonically increasing because

$$\begin{aligned} r_n &= Q_n\, Q_{n-1} \\ &= (b_n\, Q_{n-1} + Q_{n-2})\; Q_{n-1} \\ &= b_n\, Q_{n-1}^2 + r_{n-1} \\ &\geq b_n\theta^2 + r_{n-1} \end{aligned}$$

Consequently

$$r_n = (r_n - r_{n-1}) + (r_{n-1} - r_{n-2}) + \ldots + (r_1 - r_0)$$
$$\geq \theta^2(b_n + b_{n-1} + \ldots + b_1)$$

Thus if $\sum_{i=1}^{\infty} b_i$ diverges, the sequence $\{r_n\}_{n \in \mathbb{N}}$ tends to infinity. This implies, by a theorem of Leibniz on alternating series, that the Euler-Minding series

$$\sum_{i=1}^{\infty} (-1)^{i-1} \frac{1}{r_i}$$

converges. On the other hand, if $\sum_{i=1}^{\infty} b_i$ converges then we can prove that r_n is bounded above. To do so, we first prove that

$$Q_n < (1 + b_1)(1 + b_2)\ldots(1 + b_n)$$

This is obvious for $n = 1$. Assume now that it is also true for $n \leq k$, then using the recurrence relations,

$$Q_{k+1} = b_{k+1}\, Q_k + Q_{k-1}$$
$$< b_{k+1}(1 + b_1)(1 + b_2)\ldots(1 + b_k) + (1 + b_1)(1 + b_2)\ldots(1 + b_{k-1})$$
$$= (1 + b_1)\ldots(1 + b_{k-1})[b_{k+1}(1 + b_k) + 1]$$
$$< (1 + b_1)\ldots(1 + b_{k-1})(1 + b_k)(1 + b_{k+1})$$

Consequently

$$Q_n < e^{b_1}\, e^{b_2}\ldots e^{b_n}$$

because $e^x > 1 + x$ for $x > 0$.
If we put

$$\sum_{i=1}^{\infty} b_i = \tau$$

and

$$\sigma = e^\tau$$

then

$$r_n = Q_n\, Q_{n-1} < \sigma^2$$

This implies that the terms of the Euler-Minding series do not converge to zero.
∎

As an example, consider the continued fraction

$$10 + \sum_{i=1}^{\infty} \frac{2|}{|20}$$

An equivalence transformation rewrites it as

$$10 + \frac{1|}{|10} + \frac{1|}{|20} + \frac{1|}{|10} + \frac{1|}{|20} + \ldots$$

Clearly this continued fraction converges.
Note that the convergents satisfy

$$C_n - 10 = \frac{2}{20 + (C_{n-1} - 10)}$$

such that for the value C of the continued fraction

$$C = 10 + \frac{2}{10 + C}$$

or

$$C^2 = 102$$

Since all convergents are positive we get $C = \sqrt{102}$.
The next theorem is valid for continued fractions of the form (1.1.) with $b_0 = 0$, also if the partial numerators or denominators are complex numbers. It was proved by Pringsheim [18] in 1899.

Theorem 1.6.

The continued fraction

$$\sum_{i=1}^{\infty} \frac{a_i|}{|b_i}$$

converges if $|b_i| \geq |a_i| + 1 > 1$ for $i \geq 1$. For the nth convergent C_n we have $|C_n| < 1$ if $n \geq 1$.

Proof

First we prove the upper bound on C_n. Let

$$s_n(x) = \frac{a_n}{b_n + x}$$

Then

$$|s_n(0)| = \left|\frac{a_n}{b_n}\right| \leq \frac{|a_n|}{|a_n| + 1} < 1$$

Also, if $|s_{k+1} \circ \ldots \circ s_{k+n}(0)| < 1$, then

$$|s_k \circ s_{k+1} \circ \ldots \circ s_{k+n}(0)| = \left|\frac{a_k}{b_k + s_{k+1} \circ \ldots \circ s_{k+n}(0)}\right|$$
$$< \frac{|a_k|}{|b_k| - 1} \leq 1$$

Thus

$$|C_n| = |s_1 \circ s_2 \circ \ldots \circ s_n(0)| < 1$$

For the nth denominator Q_n we have

$$|Q_n| = |b_n Q_{n-1} + a_n Q_{n-2}| \geq |b_n| |Q_{n-1}| - |a_n| |Q_{n-2}|$$
$$\geq |b_n| |Q_{n-1}| - (|b_n| - 1) |Q_{n-2}|$$

and hence

$$|Q_n| - |Q_{n-1}| \geq (|b_n| - 1)(|Q_{n-1}| - |Q_{n-2}|)$$

Consequently

$$|Q_n| - |Q_{n-1}| \geq \prod_{k=1}^{n} (|b_k| - 1) \geq \prod_{k=1}^{n} |a_k| \quad \text{for} \quad n \geq 1$$

This implies that the $|Q_n|$ are monotonically increasing and that the general term in the Euler-Minding series satisfies

$$\frac{\left|\prod_{k=1}^{i} a_k\right|}{|Q_i| |Q_{i-1}|} \leq \frac{1}{|Q_{i-1}|} - \frac{1}{|Q_i|}$$

So we have

$$\sum_{i=1}^{n} \frac{\left|\prod_{k=1}^{i} a_k\right|}{|Q_i| \cdot |Q_{i-1}|} \leq \frac{1}{|Q_0|} - \frac{1}{|Q_n|} = 1 - \frac{1}{|Q_n|} \quad \text{for } n \geq 1$$

and hence the Euler-Minding series converges absolutely. This implies convergence of the given continued fraction. ∎

The earliest known convergence criterion for continued fractions with complex elements is the following result of Worpitzky [29] which can easily be proved using the preceding theorem. We don't give the original proof of 1865.

Theorem 1.7.

The continued fraction

$$\sum_{i=1}^{\infty} \frac{a_i|}{|1}$$

converges if

$$|a_i| \leq \frac{1}{4}$$

for $i \geq 2$.

Proof

Apply the previous theorem to the continued fraction

$$\frac{2a_1|}{|2} + \sum_{i=2}^{\infty} \frac{4a_i|}{|2}$$

which is equal to

$$\sum_{i=1}^{\infty} \frac{a_i|}{|1}$$

up to an equivalence transformation with $p_i = 2$ for all $i \geq 1$. ∎

An extensive treatment of the convergence problem of continued fractions is given in [11 pp. 60-146]. We also refer to the work of Wall [28] and Perron [17].

4.2. Convergence of continued fraction expansions.

Let z be a complex number. Consider the continued fraction expansion

$$b_0(z) + \sum_{i=1}^{\infty} \frac{a_i(z)|}{|b_i(z)|}$$

This expansion is only defined for those values of z for which the continued fraction converges. For some choices of $b_0(z)$, $a_i(z)$ and $b_i(z)$ a lot is known about the convergence of such a continued fraction expansion. We will give results for expansions of the following type:

$$\frac{1|}{|b_1 z} + \frac{1|}{|b_2} + \frac{1|}{|b_3 z} + \frac{1|}{|b_4} + \frac{1|}{|b_5 z} + \ldots \qquad (1.15a.)$$

and

$$\frac{a_1|}{|1} + \frac{a_2 z|}{|1} + \frac{a_3 z|}{|1} + \frac{a_4 z|}{|1} + \ldots \qquad (1.15b.)$$

A lot of continued fractions can also be written in one of the forms above by means of an equivalence transformation.

Theorem 1.8.

If for the continued fraction (1.15a.) all the b_i are real and strictly positive with $\sum_{i=1}^{\infty} b_i$ divergent, then (1.15a.) converges in every closed and bounded subset G of the complex plane for which the distance to the negative real axis is positive. The convergence is uniform to a function $f(z)$ which is holomorphic for all z not on the negative real axis.

132, 492

Domain of uniform convergence
of the continued fraction (1.15a.).

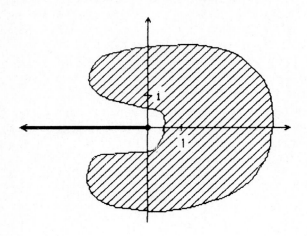

Figure 1.1.

This theorem was originally proved by Stieltjes [23, 28 p. 120] in 1894. We illustrate it with the following example. A continued fraction expansion for the function [10 p. 614]

$$\frac{2e^z}{\sqrt{z}} \int_{\sqrt{z}}^{\infty} e^{-u^2}\, du \quad (|\arg z| < \pi,\ z \neq 0)$$

is

$$\frac{1|}{|z} + \frac{1/2|}{|1} + \frac{1|}{|z} + \frac{3/2|}{|1} + \ldots$$

or after an equivalence transformation

$$\frac{1|}{|z} + \frac{1|}{|2} + \frac{1|}{|z/2} + \frac{1|}{|4/3} + \frac{1|}{|3z/4} + \frac{1|}{|8/15} + \ldots$$

According to theorem 1.8. this continued fraction expansion converges for all z not on the negative real axis. The next result is due to Van Vleck [20 p. 394] and dates from 1904.

Theorem 1.9.

If for the continued fraction (1.15b.) $\lim_{i\to\infty} a_i = 0$ with $a_i \neq 0$, then (1.15b.) converges to a meromorphic function $f(z)$. The convergence is uniform in every closed and bounded subset G of the complex plane that contains no poles of $f(z)$.

If for the continued fraction (1.15b.) $\lim_{i\to\infty} a_i = a \neq 0$, then (1.15b.) converges in the cut complex plane

$$\mathbb{C}\backslash\{z \mid z = -\frac{\lambda}{4a}, \ \lambda \geq 1\}$$

to a function $f(z)$ meromorphic in that cut complex plane.

The convergence is uniform in every closed and bounded subset G of the complex plane that contains no poles of $f(z)$ and no points of the cut

$$\{z \mid z = -\frac{\lambda}{4a}, \lambda \geq 1\}$$

Domain of uniform convergence
of the continued fraction (1.15b.)

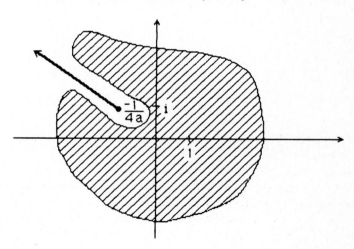

Figure 1.2.

Let us also illustrate the previous theorem with an example. A continued fraction expansion for

$$\frac{\tanh^{-1}\left(\sqrt{z}\right)}{\sqrt{z}}$$

is [1 p. 88]

$$\frac{1}{\big|1} - \frac{z}{\big|3} - \frac{4z}{\big|5} - \ldots - \frac{n^2 z}{\big|2n+1} - \ldots$$

$$= \frac{1}{\big|1} - \frac{z/3}{\big|1} - \frac{4z/15}{\big|1} - \ldots - \frac{\dfrac{n^2 z}{(2n+1)(2n-1)}}{\big|1} - \ldots$$

This expansion is of the type (1.15b.). So we calculate

$$\lim_{n\to\infty} \frac{-n^2}{(2n+1)(2n-1)} = \frac{-1}{4}$$

Theorem 1.9. says that the continued fraction expansion converges in $\mathbb{C}\backslash[1,\infty)$ and that the convergence is uniform in every closed and bounded subset of the complex plane that does not contain points of $[1,\infty)$.
The Taylor series expansion of

$$\frac{\tanh^{-1}\left(\sqrt{z}\right)}{\sqrt{z}}$$

given by

$$\frac{z}{3} + \frac{z^2}{5} + \frac{z^3}{7} + \ldots + \frac{z^n}{2n+1} + \ldots$$

only converges in the unit disc. Hence continued fraction expansions can have a larger convergence region.
A related problem is the following. If a function $f(z)$ is given, is it then possible to construct a continued fraction expansion for $f(z)$ such that the convergence region of the continued fraction expansion is exactly the domain of $f(z)$? This question has partly been answered when we discussed equivalent continued fractions. We do not study the problem in detail here but refer the interested reader to [20 pp. 386-415] and [10].

4.3. Convergence of corresponding continued fractions for Stieltjes series.

The series

$$f(z) = \sum_{i=0}^{\infty} d_i \, z^i \qquad\qquad (1.16.)$$

is called a **Stieltjes series** if

$$d_i = \int_0^{\infty} t^i dg(t)$$

where $g(t)$ is a real-valued, bounded, nondecreasing function taking on infinitely many different values. The values d_i are called **moments** of the function $g(t)$. If $g(t)$ is constant for $t > r$ with $0 < r < \infty$, then

$$d_i = \int_0^{r} t^i \, dg(t)$$

The **Stieltjes transform** of $g(t)$ is defined as

$$F(z) = \int_0^{r} \frac{1}{1 - zt} \, dg(t) \qquad z \notin [\tfrac{1}{r}, \infty) \qquad (1.17.)$$

A proof of the existence of this transform is given in [10 p. 578].
The series (1.16.) has convergence radius $1/r$ and can be regarded as a formal power series expansion of $F(z)$ which is analytic in the cut complex plane $\mathbb{C}\backslash[\tfrac{1}{r}, \infty)$ [10 p. 581].

Theorem 1.10.

The corresponding continued fraction for $f(z)$ given by (1.16.) converges to $F(z)$ given by (1.17.) for all z in $\mathbb{C}\backslash[\tfrac{1}{r}, \infty)$.
The convergence is uniform on every closed and bounded subset of the cut complex plane.

The proof which was originally given by Markov can be found in [17 p. 202].
A simple example of a Stieltjes series is

$$f(z) = \sum_{i=0}^{\infty} \frac{z^i}{i + 1}$$

Here $g(t) = t$ for $0 \leq t \leq 1$ and $g(t) = 1$ for $t \geq 1$.

The Stieltjes transform is

$$F(z) = -\frac{1}{z}\,\ln(1-z)$$

As a consequence of theorem (1.10.) we get that the corresponding continued fraction for f converges to F for all z in $\mathbb{C}\backslash[1,\infty)$.

§5. Algorithms to evaluate continued fractions.

If we want to know an approximation for the value of a continued fraction, we must compute one or more convergents C_n. The recurrence relations (1.3.) for the nth numerator and denominator provide a means to calculate the nth convergent since

$$C_n = \frac{P_n}{Q_n}$$

This algorithm is called **forward** because it is possible to compute C_{n+1} from the knowledge of C_n with little extra work. We shall now discuss some other algorithms used for the computation of convergents.

5.1. The backward algorithm.

The nth convergent C_n can easily be calculated as follows:
put

$$r_{n+1,n} = 0$$

and compute

$$r_{i,n} = \frac{a_i}{b_i + r_{i+1,n}} \qquad i = n, \ldots, 1$$

Then

$$C_n = b_0 + r_{1,n}$$

A drawback of this method is that it must fully be repeated for each convergent we want to compute. It is impossible to calculate C_{n+1} starting from C_n. But the algorithm appears to be numerically stable in a lot of cases [2].

5.2. Forward algorithms.

The following theorem can be found in [14].

Theorem 1.11.

The nth convergent of the continued fraction

$$\sum_{i=1}^{\infty} \frac{a_i|}{|b_i}$$

is the first unknown $x_{1,n}$ of the tridiagonal system of linear equations

$$\begin{pmatrix} b_1 & -1 & 0 & \cdots & 0 \\ a_2 & b_2 & -1 & & \vdots \\ 0 & a_3 & b_3 & \ddots & 0 \\ \vdots & & \ddots & \ddots & -1 \\ 0 & \cdots & 0 & a_n & b_n \end{pmatrix} \begin{pmatrix} x_{1,n} \\ \vdots \\ x_{n,n} \end{pmatrix} = \begin{pmatrix} a_1 \\ 0 \\ \vdots \\ 0 \end{pmatrix} \qquad (1.18.)$$

(see also problem (5)).

Consequently, algorithms for the solution of a linear tridiagonal system, and especially for the computation of the first unknown, are also algorithms for the calculation of C_n.

If backward Gaussian elimination is used to solve (1.18.), then the coefficient matrix of (1.18.) is transformed into a lower triangular matrix and the computation of $x_{1,n} = r_{1,n}$ is precisely the backward algorithm.

It is also easy to see that the computation of C_n via the recurrence relations (1.3.) is equivalent with solving (1.18.) by means of the so-called shooting method.

If we choose $x_{1,n} = x_{1,n}^{(0)} = 0$ then according to (1.18.) $x_{2,n}^{(0)} = -a_1$ and

$$x_{k+1,n}^{(0)} = a_k x_{k-1,n}^{(0)} + b_k x_{k,n}^{(0)} \qquad k = 2, \ldots, n-1$$

If we choose $x_{1,n} = x_{1,n}^{(1)} = 1$ then $x_{2,n}^{(1)} = b_1 - a_1$ and

$$x_{k+1,n}^{(1)} = a_k x_{k-1,n}^{(1)} + b_k x_{k,n}^{(1)} \qquad k = 2, \ldots, n-1$$

The last equation of the linear system (1.18.) is after substitution of the first $(n-1)$ equations in it, merely a linear function $g(x_{1,n})$. The $x_{1,n}$ we are looking for is a root of $g(x_{1,n})$, in other words the intersection point of the x-axis with the straight line $y = g(x)$.

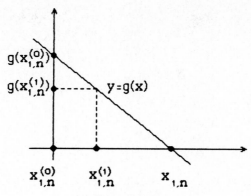

Figure 1.3.

So

$$g(x_{1,n}) - g(x_{1,n}^{(0)}) = \frac{g(x_{1,n}^{(0)}) - g(x_{1,n}^{(1)})}{x_{1,n}^{(0)} - x_{1,n}^{(1)}}\left(x_{1,n} - x_{1,n}^{(0)}\right)$$

or

$$x_{1,n} = \frac{g\left(x_{1,n}^{(0)}\right)}{g\left(x_{1,n}^{(0)}\right) - g\left(x_{1,n}^{(1)}\right)}$$

$$= \frac{b_n\, x_{n,n}^{(0)} + a_n\, x_{n-1,n}^{(0)}}{b_n\left(x_{n,n}^{(0)} - x_{n,n}^{(1)}\right) + a_n\left(x_{n-1,n}^{(0)} - x_{n-1,n}^{(1)}\right)}$$

After comparison of the starting values for the recursive formulas (1.3.), when applied to

$$\sum_{i=1}^{\infty} \left.\frac{a_i}{b_i}\right|$$

with those for the recursive computation of $x_{k,n}^{(0)}$ and $x_{k,n}^{(0)} - x_{k,n}^{(1)}$, we get

$$x_{1,n} = \frac{-P_n}{-Q_n}$$

since the recurrence relations are identical.

Another forward algorithm for the computation of C_n is obtained when (1.18.) is solved by forward Gaussian elimination and backsubstitution. The resulting formulas are

$$r_{1,n} = b_1$$

$$r_{i,n} = b_i + \frac{a_i}{r_{i-1,n}} \quad i = 2, \ldots, n$$

$$x_{1,n} = \sum_{i=1}^{n} (-1)^{i-1} \frac{a_1 \ldots a_i}{r_{1,n}^2 \ldots r_{i-1,n}^2 \ r_{i,n}}$$

$$= x_{1,n-1} - \left[(-1)^{n-2} \frac{a_1 \ldots a_{n-1}}{r_{1,n}^2 \ldots r_{n-2,n}^2 \ r_{n-1,n}} \right] \frac{a_n}{r_{n-1,n} \ r_{n,n}}$$

Finally $C_n = b_0 + x_{1,n}$.
More algorithms for the calculation of C_n can be found in the literature. We refer among others to [25] and [6].

5.3. Modifying factors.

Even efficient ways to calculate C_n do not guarantee that C_n is a good approximation for the value $C = \lim_{n \to \infty} C_n$ of the continued fraction. Since the nth convergent results from truncating

$$b_0 + \sum_{i=1}^{\infty} \frac{a_i}{\left| b_i \right.}$$

we shall call the chopped off part

$$T_n = \sum_{i=n+1}^{\infty} \frac{a_i}{\left| b_i \right.}$$

the nth **tail** of the continued fraction. Clearly

$$C = b_0 + \sum_{i=1}^{n-1} \frac{a_i}{\left| b_i \right.} + \frac{a_n}{\left| b_n + T_n \right.} \tag{1.19.}$$

and

$$T_n = \frac{a_{n+1}}{b_{n+1} + T_{n+1}}$$

Let again for $n \geq 1$

$$s_n(x) = \frac{a_n}{b_n + x}$$

Then

$$C_n = b_0 + s_1 \circ s_2 \circ \ldots \circ s_n(0)$$

and

$$C = b_0 + s_1 \circ s_2 \circ \ldots \circ s_n(T_n)$$

Hence, in order to estimate C, it may be better to replace the tail T_n by a value different from zero. In many cases the tails do not even converge to zero. Suppose τ_n is such an approximation for T_n. We shall then call

$$\Gamma_n = b_0 + s_1 \circ s_2 \circ \ldots \circ s_n(\tau_n)$$

the nth **modified convergent** with τ_n the nth **modifying factor**.
The next theorems illustrate in which cases modifying factors are really worthwile, in the sense that

$$|C - \Gamma_n| < |C - C_n|$$

First of all we study the behaviour of the tails [17 p. 93].

Theorem 1.12.

If the continued fraction (1.1.) is such that $\lim\limits_{i \to \infty} a_i = a$ and $\lim\limits_{i \to \infty} b_i = b$ with $a, b \in \mathbb{C}$ and if the quadratic equation $x^2 + bx - a = 0$ has two roots x_1 and x_2 with $|x_1| < |x_2|$ then

$$\lim_{n \to \infty} T_n = x_1$$

This behaviour of the tails suggests to choose

$$\tau_n = x_1$$

In order to study the effect of this modifying factor we rewrite the expression $|C - \Gamma_n|/|C - C_n|$ as follows. Using the three-term recurrence relation for (1.19.) we find

$$C - \Gamma_n = \frac{(b_n + T_n)P_{n-1} + a_n P_{n-2}}{(b_n + T_n)Q_{n-1} + a_n Q_{n-2}} - \frac{(b_n + \tau_n)P_{n-1} + a_n P_{n-2}}{(b_n + \tau_n)Q_{n-1} + a_n Q_{n-2}}$$

$$= \frac{P_n + T_n P_{n-1}}{Q_n + T_n Q_{n-1}} - \frac{P_n + \tau_n P_{n-1}}{Q_n + \tau_n Q_{n-1}}$$

$$= \frac{(P_n Q_{n-1} - Q_n P_{n-1})(\tau_n - T_n)}{Q_{n-1}^2 \left(T_n + \frac{Q_n}{Q_{n-1}}\right)\left(\tau_n + \frac{Q_n}{Q_{n-1}}\right)}$$

Analogously

$$C - C_n = \frac{P_n + T_n P_{n-1}}{Q_n + T_n Q_{n-1}} - \frac{P_n}{Q_n} = -T_n \frac{P_n Q_{n-1} - Q_n P_{n-1}}{Q_n Q_{n-1}\left(T_n + \frac{Q_n}{Q_{n-1}}\right)}$$

This leads to

$$\frac{C - \Gamma_n}{C - C_n} = \frac{T_n - \tau_n}{T_n} \frac{h_n}{h_n + \tau_n} = \frac{T_n - x_1}{T_n} \frac{h_n}{h_n + x_1}$$

with

$$h_n = \frac{Q_n}{Q_{n-1}} = b_n + \frac{a_n}{h_{n-1}}$$

where

$$h_0 = \infty$$

If the continued fraction (1.1.) satisfies the conditions of theorem 1.12. then we shall denote

$$D = |b + x_1| - |x_1|$$

$$d_n = \max_{m \geq n} |a_m - a|$$

$$e_n = \max_{m \geq n} |b_m - b|$$

$$E_n = \frac{d_n D}{D^2 - 2d_n}$$

$$a_n = a + \alpha_n$$

$$b_n = b + \beta_n$$

$$T_n = x_1 + \xi_n$$

$$\theta_n = h_n + x_1$$

Using these notations we can formulate the next theorem [24].

Theorem 1.13.

Let the continued fraction

$$b_0 + \sum_{i=1}^{\infty} \frac{a_i|}{|b_i}$$

be such that $\lim_{i \to \infty} a_i = a$ and $\lim_{i \to \infty} b_i = b$ with $a, b \in \mathbb{C}$ and let x_1 be the strictly smallest root of the quadratic equation $x^2 + bx - a = 0$ with

$$|b + x_1| - |x_1| > 0$$

If also

$$d_n \leq \min \left(\frac{D^2}{3}, \frac{|a|}{2} \right) \qquad (1.20a.)$$

$$d_n + e_n \left(|b + x_1| - E_n \right) \leq \frac{d_n D^2 (D^2 - 3d_n)}{(D^2 - 2d_n)^2} \qquad (1.20b.)$$

$$e_n + d_n \left| \frac{Q_{n-2}}{Q_{n-1}} \right| \leq \frac{D^2}{2 \left(|b + x_1| + |x_1| \right)} \qquad (1.20c.)$$

then

$$|C - \Gamma_n| \leq 6d_n \frac{|b + x_1| + |x_1|}{|x_1| \left(|b + x_1| - |x_1| \right)^2} |C - C_n|$$

Proof

Let us first show that in (1.20b.) we have $|b + x_1| \geq E_n$. We know that $d_n \leq D^2/3$ and this implies

$$D^2 \geq \left(\frac{|b + x_1| - |x_1|}{|b + x_1|} + 2 \right) d_n$$

since

$$0 \leq \frac{|b + x_1| - |x_1|}{|b + x_1|} \leq 1$$

Hence

$$(D^2 - 2d_n)|b + x_1| \geq D d_n$$

or

$$|b + x_1| \geq E_n$$

In order to bound $|(C - \Gamma_n)/(C - C_n)|$ we shall now calculate an upper bound for $|(T_n - x_1)/T_n|$. Since

$$\lim_{n \to \infty} T_n = x_1$$

and

$$\lim_{n \to \infty} b_n = b$$

we have

$$\lim_{n \to \infty} \xi_n = 0$$

and

$$\lim_{n \to \infty} \beta_n = 0$$

and hence for fixed n

$$\exists m, \forall k \geq m : |\xi_k + \beta_k| \leq E_n$$

We can then write for fixed $k > \max(m, n)$

$$
\begin{aligned}
\xi_{k-1} = T_{k-1} - x_1 &= \frac{a + \alpha_k}{b + \beta_k + T_k} - x_1 \\
&= \frac{a + \alpha_k - x_1(b + \beta_k + x_1 + \xi_k)}{b + \beta_k + x_1 + \xi_k} \\
&= \frac{a - x_1 b - x_1^2 + \alpha_k - x_1(\beta_k + \xi_k)}{b + \beta_k + x_1 + \xi_k} \\
&= \frac{\alpha_k - x_1(\beta_k + \xi_k)}{(b + x_1) + (\beta_k + \xi_k)}
\end{aligned}
$$

and

$$|\xi_{k-1} + \beta_{k-1}| \le \frac{|\alpha_k| + |x_1|E_n}{|b + x_1| - E_n} + |\beta_{k-1}| \le E_n$$

because

$$|\alpha_k| + |\beta_{k-1}|(|b + x_1| - E_n) \le d_n + e_n(|b + x_1| - E_n)$$
$$\le DE_n - E_n^2$$

Using this upper bound for $|\xi_{k-1} + \beta_{k-1}|$ we can also prove it to be an upper bound for $|\xi_{k-2} + \beta_{k-2}|$. Repeating this procedure as long as $|\alpha_k| \le d_n$ and $|\beta_{k-1}| \le e_n$, finally assures for $k - 1 = n$

$$|\xi_n + \beta_n| \le E_n$$

and consequently

$$\left| \frac{T_n - x_1}{T_n} \right| = \left| \frac{a + \alpha_{n+1}}{b + \beta_{n+1} + x_1 + \xi_{n+1}} - x_1 \right| \left| \frac{b + \beta_{n+1} + x_1 + \xi_{n+1}}{a + \alpha_{n+1}} \right|$$

$$= \left| \frac{\alpha_{n+1} - x_1(\beta_{n+1} + \xi_{n+1})}{(b + x_1)x_1 + \alpha_{n+1}} \right|$$

$$\le \frac{d_n + |x_1|E_n}{|(b + x_1)x_1| - d_n}$$

$$= \frac{d_n}{|(b + x_1)x_1| - d_n} \frac{D^2 - 2d_n + |x_1|D}{D^2 - 2d_n}$$

Now since $d_n \le |a|/2 = |(b + x_1)x_1|/2$ and $2d_n \le 2D^2/3$ we have

$$\left| \frac{T_n - x_1}{T_n} \right| \le \frac{2d_n}{|(b + x_1)x_1|} \frac{3(D^2 + |x_1|D)}{D^2}$$

$$= \frac{6d_n(D + |x_1|)}{|b + x_1||x_1|D}$$

$$= \frac{6d_n}{|x_1|D}$$

Next we shall compute an upper bound for $|h_n/(h_n + x_1)|$ which is the second factor in $|(C - \Gamma_n)/(C - C_n)|$. Note already that

$$|\theta_0| = |h_0 + x_1| = \infty \ge \frac{D}{2}$$

By induction it follows that

$$|\theta_n| = |h_n + x_1| = \left| b + \beta_n + \frac{a + \alpha_n}{\theta_{n-1} - x_1} + x_1 \right|$$

$$= \left| \frac{(b + x_1)\theta_{n-1} + \alpha_n + \beta_n(\theta_{n-1} - x_1)}{\theta_{n-1} - x_1} \right|$$

$$\geq \left| \frac{b + x_1}{1 - \frac{x_1}{\theta_{n-1}}} \right| - \left| \beta_n + \frac{\alpha_n}{\theta_{n-1} - x_1} \right|$$

$$\geq \frac{|b + x_1|}{1 + \frac{2|x_1|}{D}} - \left(e_n + d_n \left| \frac{Q_{n-2}}{Q_{n-1}} \right| \right) \geq \frac{D}{2}$$

because

$$e_n + d_n \left| \frac{Q_{n-2}}{Q_{n-1}} \right| \leq \frac{D^2}{2(|b + x_1| + |x_1|)}$$

This gives us

$$\left| \frac{h_n}{h_n + x_1} \right| = \left| \frac{\theta_n - x_1}{\theta_n} \right| \leq 1 + \left| \frac{x_1}{\theta_n} \right| \leq \frac{|b + x_1| + |x_1|}{D}$$

Using the estimates for $|(T_n - x_1)/T_n|$ and $|h_n/(h_n + x_1)|$ it is now easy to finish the proof. We have

$$\left| \frac{C - \Gamma_n}{C - C_n} \right| \leq \left| \frac{T_n - x_1}{T_n} \right| \left| \frac{h_n}{h_n + x_1} \right|$$

$$\leq \frac{6d_n}{|x_1|D} \frac{|b + x_1| + |x_1|}{D}$$

$$= 6d_n \frac{|b + x_1| + |x_1|}{|x_1|(|b + x_1| - |x_1|)^2} \qquad \blacksquare$$

§6. Branched continued fractions.

6.1. Definition of branched continued fractions.

If the denominators b_i in the continued fraction

$$b_0 + \sum_{i=1}^{\infty} \frac{a_i}{\left| b_i \right.}$$

are themselves infinite expressions, then it is called a **branched continued fraction**. The b_i are called the branches and we need a multi-index to indicate a convergent. Consider for instance the expression

$$\sum_{j=0}^{\infty} \frac{a_j^{(0)}}{\left| b_j^{(0)} \right.} + \sum_{i=1}^{\infty} \frac{a_i}{\left| \displaystyle\sum_{j=0}^{\infty} \frac{a_j^{(i)}}{\left| b_j^{(i)} \right.} \right.}$$

The $(n, m_0, m_1, \ldots, m_n)^{th}$ **convergent** is then the subexpression

$$C_{n,m_0,m_1,\ldots,m_n} = \sum_{j=0}^{m_0} \frac{a_j^{(0)}}{\left| b_j^{(0)} \right.} + \sum_{i=1}^{n} \frac{a_i}{\left| \displaystyle\sum_{j=0}^{m_i} \frac{a_j^{(i)}}{\left| b_j^{(i)} \right.} \right.}$$

We will use branched continued fractions to construct a multivariate Viscovatov algorithm for the computation of multivariate continued fraction expansions of the form

$$b_{0,0}^{(0)} + \sum_{j=0}^{\infty} \frac{a_{1,j}^{(0)}}{\left| b_{1,j}^{(0)} \right.} + \ldots + \sum_{j=0}^{\infty} \frac{a_{k,j}^{(0)}}{\left| b_{k,j}^{(0)} \right.} + \sum_{i=1}^{\infty} \frac{a_i}{\left| b_{0,0}^{(i)} + \displaystyle\sum_{j=0}^{\infty} \frac{a_{1,j}^{(i)}}{\left| b_{1,j}^{(i)} \right.} + \ldots + \displaystyle\sum_{j=0}^{\infty} \frac{a_{k,j}^{(i)}}{\left| b_{k,j}^{(i)} \right.} \right.}$$

where k is the number of variables we are dealing with. Input of such an algorithm is a multivariate power series.
Vice versa, given a branched continued fraction, we can also construct an Euler-Minding series of which the successive partial sums equal a given sequence of convergents.

6.2. A generalization of the Euler-Minding series.

Let us consider continued fractions

$$B_i = b_0^{(i)} + \frac{a_1^{(i)}}{\left| b_1^{(i)} \right.} + \frac{a_2^{(i)}}{\left| b_2^{(i)} \right.} + \dots \qquad (1.21a.)$$

for $i = 0, 1, 2, \dots$.

If $C_n^{(i)}$ denotes the nth convergent of (1.21a.) then according to (1.4.)

$$C_n^{(i)} - C_{n-1}^{(i)} = (-1)^{(n+1)} \frac{a_1^{(i)} \dots a_n^{(i)}}{Q_n^{(i)} \, Q_{n-1}^{(i)}} \qquad n = 1, 2, \dots$$

$$C_0^{(i)} = b_0^{(i)}$$

where we have written

$$C_n^{(i)} = \frac{P_n^{(i)}}{Q_n^{(i)}}$$

We will now generalize (1.4.) for the branched continued fraction

$$B_0 + \frac{a_1}{\left| B_1 \right.} + \frac{a_2}{\left| B_2 \right.} + \dots \qquad (1.21b.)$$

Let us denote by P_n/Q_n the subexpression

$$C_{n,n-1,\dots,1,0}$$

$$= b_0^{(0)} + \sum_{j=1}^{n} \frac{a_j^{(0)}}{\left| b_j^{(0)} \right.} + \sum_{i=1}^{n} \frac{a_i}{\left| b_0^{(i)} + \sum_{j=1}^{n-i} \frac{a_j^{(i)}}{\left| b_j^{(i)} \right.} \right.}$$

$$= C_n^{(0)} + \sum_{i=1}^{n} \frac{a_i}{\left| C_{n-i}^{(i)} \right.} \qquad (1.22.)$$

So P_n/Q_n is the $(n, n, n-1, \dots, 1, 0)^{th}$ convergent of (1.20.). Another subexpression we shall need is

$$\frac{R_k^{(n)}}{S_k^{(n)}} = C_n^{(0)} + \sum_{i=1}^{k} \frac{a_i}{\left| C_{n-i}^{(i)} \right.} \qquad k = 0, \ldots, n \qquad (1.23.)$$

which is in fact the kth convergent of P_n/Q_n. These subconvergents can be ordered in a table

$$\frac{R_0^{(0)}}{S_0^{(0)}}$$

$$\frac{R_0^{(1)}}{S_0^{(1)}} \qquad \frac{R_1^{(1)}}{S_1^{(1)}}$$

$$\frac{R_0^{(2)}}{S_0^{(2)}} \qquad \frac{R_1^{(2)}}{S_1^{(2)}} \qquad \frac{R_2^{(2)}}{S_2^{(2)}}$$

$$\vdots \qquad \vdots \qquad \vdots \qquad \ddots$$

where we proceed in a certain row from one value to the next one by using (1.3.) for (1.22.) :

$$R_k^{(n)} = C_{n-k}^{(k)} R_{k-1}^{(n)} + a_k R_{k-2}^{(n)}$$

$$\qquad\qquad\qquad\qquad k = 1, \ldots, n \qquad (1.24.)$$

$$S_k^{(n)} = C_{n-k}^{(k)} S_{k-1}^{(n)} + a_k S_{k-2}^{(n)}$$

with $R_{-1}^{(n)} = 1 = S_0^{(n)}$, $R_0^{(n)} = C_n^{(0)}$ and $S_{-1}^{(n)} = 0$. If we want to develop a formula analogous to (1.5.) for the branched continued fraction (1.21.) we must compute an expression for the difference

$$\frac{P_n}{Q_n} - \frac{P_{n-1}}{Q_{n-1}} = \frac{R_n^{(n)}}{S_n^{(n)}} - \frac{R_{n-1}^{(n-1)}}{S_{n-1}^{(n-1)}} \qquad (1.25.)$$

Remark that in comparison with P_{n-1}/Q_{n-1} the expression P_n/Q_n contains an extra term in each of the involved convergents of B_i. Also B_n is not taken into account in P_{n-1}/Q_{n-1}. In order to compute (1.25.) we must be able to proceed from one row in the table of subconvergents to the next row. The following theorem is a means to calculate the differences $R_k^{(n)} - R_k^{(n-1)}$ and $S_k^{(n)} - S_k^{(n-1)}$.

Theorem 1.14.

For $n \geq 2$ and $k = 1, \ldots, n-1$

$$R_k^{(n)} - R_k^{(n-1)} = C_{n-k}^{(k)}(R_{k-1}^{(n)} - R_{k-1}^{(n-1)}) + a_k(R_{k-2}^{(n)} - R_{k-2}^{(n-1)})$$

$$+ (-1)^{n-k+1} \frac{a_1^{(k)} \ldots a_{n-k}^{(k)}}{Q_{n-k}^{(k)} Q_{n-k-1}^{(k)}} R_{k-1}^{(n-1)}$$

$$S_k^{(n)} - S_k^{(n-1)} = C_{n-k}^{(k)}(S_{k-1}^{(n)} - S_{k-1}^{(n-1)}) + a_k(S_{k-2}^{(n)} - S_{k-2}^{(n-1)})$$

$$+ (-1)^{n-k+1} \frac{a_1^{(k)} \ldots a_{n-k}^{(k)}}{Q_{n-k}^{(k)} Q_{n-k-1}^{(k)}} S_{k-1}^{(n-1)}$$

with

$$R_{-1}^{(n)} - R_{-1}^{(n-1)} = S_{-1}^{(n)} - S_{-1}^{(n-1)} = S_0^{(n)} - S_0^{(n-1)} = 0$$

and

$$R_0^{(n)} - R_0^{(n-1)} = (-1)^{n+1} \frac{a_1^{(0)} \ldots a_n^{(0)}}{Q_n^{(0)} Q_{n-1}^{(0)}}$$

Proof

We shall perform the proof only for $R_k^{(n)} - R_k^{(n-1)}$ because it is completely analogous for $S_k^{(n)} - S_k^{(n-1)}$. Choose k and n and write down the recurrence relation (1.24.) for row n and row $n-1$ in the table of subconvergents:

$$R_k^{(n)} = C_{n-k}^{(k)} R_{k-1}^{(n)} + a_k R_{k-2}^{(n)}$$

$$R_k^{(n-1)} = C_{n-1-k}^{(k)} R_{k-1}^{(n-1)} + a_k R_{k-2}^{(n-1)}$$

By subtracting we get

$$R_k^{(n)} - R_k^{(n-1)} = C_{n-k}^{(k)}(R_{k-1}^{(n)} - R_{k-1}^{(n-1)}) + a_k(R_{k-2}^{(n)} - R_{k-2}^{(n-1)}) +$$

$$+ (C_{n-k}^{(k)} - C_{n-k-1}^{(k)}) R_{k-1}^{(n-1)}$$

where by (1.4.)

$$C_{n-k}^{(k)} - C_{n-k-1}^{(k)} = (-1)^{n-k+1} \frac{a_1^{(k)} \ldots a_{n-k}^{(k)}}{Q_{n-k}^{(k)} Q_{n-k-1}^{(k)}}$$

The first three starting starting values are easy to check and for $R_0^{(n)} - R_0^{(n-1)}$ again (1.4.) is used. ∎

From the above theorem we see that up to an additional correction term the values $R_k^{(n)} - R_k^{(n-1)}$ and $S_k^{(n)} - S_k^{(n-1)}$ also satisfy a three-term recurrence relation. By means of this result we can write for the numerator of (1.25.):

$$R_n^{(n)} S_{n-1}^{(n-1)} - S_n^{(n)} R_{n-1}^{(n-1)} = \left(C_0^{(n)} R_{n-1}^{(n)} + a_n R_{n-2}^{(n)} \right) S_{n-1}^{(n-1)}$$

$$- \left(C_0^{(n)} S_{n-1}^{(n)} + a_n S_{n-2}^{(n)} \right) R_{n-1}^{(n-1)}$$

$$= C_0^{(n)} S_{n-1}^{(n-1)} \left[R_{n-1}^{(n-1)} + C_1^{(n-1)} \left(R_{n-2}^{(n)} - R_{n-2}^{(n-1)} \right) + \right.$$

$$\left. + a_{n-1} \left(R_{n-3}^{(n)} - R_{n-3}^{(n-1)} \right) + \frac{a_1^{(n-1)}}{b_1^{(n-1)}} R_{n-2}^{(n-1)} \right]$$

$$- C_0^{(n)} R_{n-1}^{(n-1)} \left[S_{n-1}^{(n-1)} + C_1^{(n-1)} \left(S_{n-2}^{(n)} - S_{n-2}^{(n-1)} \right) + \right.$$

$$\left. + a_{n-1} \left(S_{n-3}^{(n)} - S_{n-3}^{(n-1)} \right) + \frac{a_1^{(n-1)}}{b_1^{(n-1)}} S_{n-2}^{(n-1)} \right]$$

$$+ a_n \left(R_{n-2}^{(n)} S_{n-1}^{(n-1)} - S_{n-2}^{(n)} R_{n-1}^{(n-1)} \right)$$

$$= \left(C_0^{(n)} C_1^{(n-1)} + a_n \right) \left[S_{n-1}^{(n-1)} \left(R_{n-2}^{(n)} - R_{n-2}^{(n-1)} \right) - R_{n-1}^{(n-1)} \left(S_{n-2}^{(n)} - S_{n-2}^{(n-1)} \right) \right]$$

$$+ C_0^{(n)} a_{n-1} \left[S_{n-1}^{(n-1)} \left(R_{n-3}^{(n)} - R_{n-3}^{(n-1)} \right) - R_{n-1}^{(n-1)} \left(S_{n-3}^{(n)} - S_{n-3}^{(n-1)} \right) \right]$$

$$+ \left(C_0^{(n)} \frac{a_1^{(n-1)}}{b_1^{(n-1)}} + a_n \right) \left[R_{n-2}^{(n-1)} S_{n-1}^{(n-1)} - S_{n-2}^{(n-1)} R_{n-1}^{(n-1)} \right]$$

where

$$R_{n-2}^{(n-1)} S_{n-1}^{(n-1)} - S_{n-2}^{(n-1)} R_{n-1}^{(n-1)} = (-1)^{n-1} a_1 \ldots a_{n-1}$$

because $R_{n-2}^{(n-1)}/S_{n-2}^{(n-1)}$ and $R_{n-1}^{(n-1)}/S_{n-1}^{(n-1)}$ are consecutive convergents of the finite continued fraction

$$C_{n-1}^{(0)} + \sum_{i=1}^{n-1} \frac{a_i}{\left| C_{n-1-i}^{(i)} \right.}$$

In this way

$$\frac{P_n}{Q_n} - \frac{P_{n-1}}{Q_{n-1}} =$$

$$\frac{(C_0^{(n)} C_1^{(n-1)} + a_n)(R_{n-2}^{(n)} - R_{n-2}^{(n-1)}) + C_0^{(n)} a_{n-1}(R_{n-3}^{(n)} - R_{n-3}^{(n-1)})}{Q_n}$$

$$- \frac{P_{n-1}}{Q_n Q_{n-1}} \left[(C_0^{(n)} C_1^{(n-1)} + a_n)(S_{n-2}^{(n)} - S_{n-2}^{(n-1)}) + \right.$$

$$\left. + C_0^{(n)} a_{n-1}(S_{n-3}^{(n)} - S_{n-3}^{(n-1)}) \right]$$

$$+ (-1)^{n-1} \frac{a_1 \ldots a_{n-1}}{Q_n Q_{n-1}} \left(C_0^{(n)} \frac{a_1^{(n-1)}}{b_1^{(n-1)}} + a_n \right) \tag{1.26.}$$

We remark that (1.26.) reduces to (1.4.) if the continued fraction (1.21.) is not branched because then $R_k^{(k)} = R_k^{(n)}$ and $S_k^{(k)} = S_k^{(n)}$ for all $n \geq k$. Consequently the classical Euler-Minding series will turn out to be a special case of the Euler-Minding series for branched continued fractions.

Theorem 1.15.

For $n \geq 2$ the convergent $C_{n,n,n-1,\ldots,1,0}$ of the branched continued fraction (1.21.) can be written as

$$\frac{P_n}{Q_n} = C_1^{(0)} + \frac{a_1}{C_0^{(1)}} + \sum_{i=2}^{n} (-1)^{i+1} \frac{a_1 \ldots a_{i-1}}{Q_i Q_{i-1}} \left(a_i + C_0^{(i)} \frac{a_1^{(i-1)}}{b_1^{(i-1)}} \right)$$

$$+ \sum_{i=2}^{n} \frac{(a_i + C_0^{(i)} C_1^{(i-1)})(R_{i-2}^{(i)} - R_{i-2}^{(i-1)}) + C_0^{(i)} a_{i-1}(R_{i-3}^{(i)} - R_{i-3}^{(i-1)})}{Q_i}$$

$$- \sum_{i=2}^{n} \frac{P_{i-1}}{Q_{i-1}} \frac{(a_i + C_0^{(i)} C_1^{(i-1)})(S_{i-2}^{(i)} - S_{i-2}^{(i-1)}) + C_0^{(i)} a_{i-1}(S_{i-3}^{(i)} - S_{i-3}^{(i-1)})}{Q_i}$$

Proof

The result is obvious if we write

$$\frac{P_n}{Q_n} = \frac{P_1}{Q_1} + \sum_{i=2}^{n} \left(\frac{P_i}{Q_i} - \frac{P_{i-1}}{Q_{i-1}} \right)$$

and insert (1.26.) for $P_i/Q_i - P_{i-1}/Q_{i-1}$. ∎

As a result of the previous theorem we can associate with the branched continued fraction (1.21.) the series

$$C_1^{(0)} + \frac{a_1}{C_0^{(1)}} + \sum_{i=2}^{\infty} \left\{ (-1)^{i+1} \frac{a_1 \ldots a_{i-1}}{Q_i Q_{i-1}} \left(a_i + C_0^{(i)} \frac{a_1^{(i-1)}}{b_1^{(i-1)}} \right) + \right.$$

$$+ \frac{(a_i + C_0^{(i)} C_1^{(i-1)})(R_{i-2}^{(i)} - R_{i-2}^{(i-1)}) + C_0^{(i)} a_{i-1}(R_{i-3}^{(i)} - R_{i-3}^{(i-1)})}{Q_i} +$$

$$\left. - \frac{P_{i-1}}{Q_{i-1}} \frac{(a_i + C_0^{(i)} C_1^{(i-1)})(S_{i-2}^{(i)} - S_{i-2}^{(i-1)}) + C_0^{(i)} a_{i-1}(S_{i-3}^{(i)} - S_{i-3}^{(i-1)})}{Q_i} \right\}$$

of which the successive partial sums equal the successive convergents $C_{n,n,n-1,\ldots,1,0}$ of (1.21.).

6.3. Some recurrence relations.

In order to formulate a multivariate Viscovatov algorithm we first show that a corresponding continued fraction can be obtained from a system of recurrence relations [16].
Consider the problem of constructing a continued fraction expansion of the form

$$f(x) = \frac{a_1}{|1|} + \frac{a_2 x}{|1|} + \frac{a_3 x}{|1|} + \ldots \qquad (1.27.)$$

for a given series expansion

$$f(x) = c_1^{(0)} + c_2^{(0)} x + c_3^{(0)} x^2 + \ldots$$

Remark that (1.27.) coincides with (1.14.) after an equivalence transformation. Instead of using Viscovatov's algorithm, the coefficients a_i can also be deduced from the following set of recurrence relations. Define

$$f_0 = f \quad \text{given by} \quad (1.27.)$$
$$f_1 = a_1 - f_0 \qquad\qquad\qquad (1.28a.)$$
$$f_k = a_k x f_{k-2} - f_{k-1} \qquad k = 2, 3, 4, \ldots \qquad (1.28b.)$$

As mentioned in section 3.4. $f_k(x)$ can be developed in a series of the form

$$f_k(x) = x^k \sum_{i=1}^{\infty} c_i^{(k)} x^{i-1}$$

Equating coefficients in relation (1.28a.)

$$x \sum_{i=1}^{\infty} c_i^{(1)} x^{i-1} = a_1 - \sum_{i=1}^{\infty} c_i^{(0)} x^{i-1}$$

we find

$$a_1 = c_1^{(0)}$$
$$c_i^{(1)} = -c_{i+1}^{(0)} \qquad i \geq 1$$

and for $k \geq 2$ by means of (1.28b.)

$$x^k \sum_{i=1}^{\infty} c_i^{(k)} x^{i-1} = a_k x^{k-1} \sum_{i=1}^{\infty} c_i^{(k-2)} x^{i-1} - x^{k-1} \sum_{i=1}^{\infty} c_i^{(k-1)} x^{i-1}$$

we get

$$a_k = \frac{c_1^{(k-1)}}{c_1^{(k-2)}}$$
$$c_i^{(k)} = a_k \ c_{i+1}^{(k-2)} - c_{i+1}^{(k-1)} \qquad i \geq 1$$

Using these formulas all the coefficients in the continued fraction (1.27.) can be computed. Hence we can also construct a continued fraction expansion of the form

$$c_0^{(0)} + \sum_{i=1}^{\infty} \frac{a_i x}{1}$$

for the power series $f(x) = c_0^{(0)} + c_1^{(0)} x + c_2^{(0)} x^2 + \ldots$ by applying the previous reasoning to $(f(x) - c_0^{(0)})/x$.

6.4. A multivariate Viscovatov algorithm.

Let us now apply this reasoning to the following problem [15]. We restrict ourselves to the bivariate case only to avoid notational difficulties. Given a double power series

$$f(x,y) = \sum_{i,j=1}^{\infty} c_{ij}^{(0)} \; x^{i-1} y^{j-1}$$

try to find a branched continued fraction of the form

$$f(x,y) = \left.\frac{a_{11}}{\left|1 + g_1(x) + h_1(y)\right.}\right| + \left.\frac{a_{22}xy}{\left|1 + g_2(x) + h_2(y)\right.}\right| + \left.\frac{a_{33}xy}{\left|1 + g_3(x) + h_3(y)\right.}\right| + \ldots \tag{1.29.}$$

with

$$g_k(x) = \frac{a_{k+1,k}\,x}{\left|\,1\,\right|} + \frac{a_{k+2,k}\,x}{\left|\,1\,\right|} + \ldots$$

$$k \geq 1$$

$$h_k(y) = \frac{a_{k,k+1}\,y}{\left|\,1\,\right|} + \frac{a_{k,k+2}\,y}{\left|\,1\,\right|} + \ldots$$

We define

$$f_0 = f \quad \text{given by} \quad (1.29.)$$
$$f_1 = a_{11} - (1 + g_1 + h_1)f_0 \tag{1.30a.}$$
$$f_k = a_{kk}\,xy\,f_{k-2} - (1 + g_k + h_k)f_{k-1} \qquad k = 2, 3, 4, \ldots \tag{1.30b.}$$

A series expansion for $f_k(x,y)$ is then of the form

$$f_k(x,y) = (xy)^k \sum_{i,j=1}^{\infty} c_{ij}^{(k)} \, x^{i-1} y^{j-1}$$

while $g_k(x)$ and $h_k(y)$ can be written as

$$g_k(x) = x \sum_{i=1}^{\infty} d_i^{(k)} x^{i-1} \qquad (1.31.)$$

$$h_k(y) = y \sum_{j=1}^{\infty} e_j^{(k)} y^{j-1} \qquad (1.32.)$$

Equating coefficients in formula (1.30a.)

$$xy \sum_{i,j=1}^{\infty} c_{ij}^{(1)} x^{i-1} y^{j-1} =$$

$$a_{11} - \left(1 + x \sum_{i=1}^{\infty} d_i^{(1)} x^{i-1} + y \sum_{j=1}^{\infty} e_j^{(1)} y^{j-1} \right) \sum_{i,j=1}^{\infty} c_{ij}^{(0)} x^{i-1} y^{j-1}$$

we obtain

$$a_{11} = c_{11}^{(0)}$$

$$d_i^{(1)} = \frac{1}{c_{11}^{(0)}} \left(-c_{i+1,1}^{(0)} - \sum_{\ell=1}^{i-1} d_\ell^{(1)} c_{i+1-\ell,1}^{(0)} \right) \qquad i \geq 1$$

$$e_j^{(1)} = \frac{1}{c_{11}^{(0)}} \left(-c_{1,j+1}^{(0)} - \sum_{\ell=1}^{j-1} e_\ell^{(1)} c_{1,j+1-\ell}^{(0)} \right) \qquad j \geq 1$$

$$c_{ij}^{(1)} = -c_{i+1,j+1}^{(0)} - \sum_{\ell=1}^{i} d_\ell^{(1)} c_{i+1-\ell,j+1}^{(0)} - \sum_{\ell=1}^{j} e_\ell^{(1)} c_{i+1,j+1-\ell}^{(0)} \quad i,j \geq 1$$

and doing the same with (1.30b.)

$$(xy)^k \sum_{i,j=1}^{\infty} c_{ij}^{(k)} x^{i-1} y^{j-1} = a_{kk}(xy)^{k-1} \sum_{i,j=1}^{\infty} c_{ij}^{(k-2)} x^{i-1} y^{j-1}$$

$$- \left(1 + x \sum_{i=1}^{\infty} d_i^{(k)} x^{i-1} + y \sum_{j=1}^{\infty} e_j^{(k)} y^{j-1} \right) (xy)^{k-1} \sum_{i,j=1}^{\infty} c_{ij}^{(k-1)} x^{i-1} y^{j-1}$$

we find for $k \geq 2$

$$a_{kk} = \frac{c_{11}^{(k-1)}}{c_{11}^{(k-2)}}$$

$$d_i^{(k)} = \frac{1}{c_{11}^{(k-1)}} \left(a_{kk} c_{i+1,1}^{(k-2)} - c_{i+1,1}^{(k-1)} - \sum_{\ell=1}^{i-1} d_\ell^{(k)} c_{i+1-\ell,1}^{(k-1)} \right) \qquad i \geq 1$$

$$e_j^{(k)} = \frac{1}{c_{11}^{(k-1)}} \left(a_{kk} c_{1,j+1}^{(k-2)} - c_{1,j+1}^{(k-1)} - \sum_{\ell=1}^{j-1} e_\ell^{(k)} c_{1,j+1-\ell}^{(k-1)} \right) \qquad j \geq 1$$

$$c_{ij}^{(k)} = a_{kk} c_{i+1,j+1}^{(k-2)} - c_{i+1,j+1}^{(k-1)} - \sum_{\ell=1}^{i} d_\ell^{(k)} c_{i+1-\ell,j+1}^{(k-1)} - \sum_{\ell=1}^{j} e_\ell^{(k)} c_{i+1,j+1-\ell}^{(k-1)}$$

$$i, j \geq 1$$

The coefficients $a_{k+i,k}$ and $a_{k,k+j}$ are computed by applying (1.27.) and (1.28.) to the series (1.31.) and (1.32.)
As a consequence one can obtain a continued fraction expansion of the form

$$c_{00}^{(0)} + \sum_{i=1}^{\infty} \frac{a_{i0} x}{1} + \sum_{j=1}^{\infty} \frac{a_{0j} y}{1} + \sum_{k=1}^{\infty} \frac{a_{kk} xy}{1 + g_k(x) + h_k(y)}$$

for a double power series

$$f(x, y) = \sum_{i,j=0}^{\infty} c_{ij}^{(0)} x^i y^j$$

by applying the previous reasoning to the power series

$$\frac{f(x, y) - c_{00}^{(0)} - \sum_{i=1}^{\infty} c_{i0}^{(0)} x^i - \sum_{j=1}^{\infty} c_{0j}^{(0)} y^j}{xy}$$

and compute the coefficients a_{i0} and a_{0j} in

$$c_{00}^{(0)} + g_0(x) + h_0(y) = c_{00}^{(0)} + \sum_{i=1}^{\infty} \frac{a_{i0} x}{1} + \sum_{j=1}^{\infty} \frac{a_{0j} y}{1}$$

by applying (1.27.) and (1.28.) to the series

$$\sum_{i=1}^{\infty} c_{i0}^{(0)} x^i$$

and

$$\sum_{j=1}^{\infty} c_{0j}^{(0)} y^j$$

To illustrate this technique we consider the following example. Take

$$f(x,y) = e^{x+y}$$

$$= 1 + x + y + \frac{1}{2}x^2 + xy + \frac{1}{2}y^2 + \frac{1}{6}x^3 + \frac{1}{2}x^2y + \frac{1}{2}xy^2 + \frac{1}{6}y^3$$

$$+ \frac{1}{24}x^4 + \frac{1}{6}x^3y + \frac{1}{4}x^2y^2 + \frac{1}{6}xy^3 + \frac{1}{24}y^4 + \ldots$$

$$= 1 + x(1 + \frac{1}{2}x + \frac{1}{6}x^2 + \ldots) + y(1 + \frac{1}{2}y + \frac{1}{6}y^2 + \ldots)$$

$$+ xy(1 + \frac{1}{2}x + \frac{1}{2}y + \frac{1}{6}x^2 + \frac{1}{4}xy + \frac{1}{6}y^2 + \ldots)$$

Since the problem is completely symmetric it is sufficient to calculate the coefficients a_{i0}, $d_i^{(k)}$ and $c_{ij}^{(k)}$ with $i \le j$. Then

$$a_{0j} = a_{j0}$$
$$e_j^{(k)} = d_j^{(k)}$$
$$c_{ji}^{(k)} = c_{ij}^{(k)} \qquad i \le j$$

Using the above formulas we obtain

$$a_{11} = 1$$

$$d_1^{(1)} = \frac{-1}{2} \quad d_2^{(1)} = \frac{1}{12} \quad d_3^{(1)} = 0 \quad \ldots$$

$$c_{11}^{(1)} = \frac{1}{4}$$

$$c_{21}^{(1)} = \frac{1}{12} \quad c_{12}^{(1)} = \frac{1}{12}$$

$$c_{31}^{(1)} = \frac{1}{48} \quad c_{22}^{(1)} = \frac{1}{36} \quad c_{13}^{(1)} = \frac{1}{48}$$

and

$$a_{22} = \tfrac{1}{4}$$

$$d_1^{(2)} = \tfrac{1}{6} \quad d_2^{(2)} = \tfrac{1}{36} \quad \dots$$

Applying (1.27.) and (1.28.) to

$$\frac{g_1(x)}{x} = -\frac{1}{2} + \frac{1}{12}x + 0x^2 + \dots$$

$$\frac{h_1(y)}{y} = -\frac{1}{2} + \frac{1}{12}y + 0y^2 + \dots$$

and

$$\frac{g_2(x)}{x} = \frac{1}{6} + \frac{1}{36}x + \dots$$

$$\frac{h_2(y)}{y} = \frac{1}{6} + \frac{1}{36}y + \dots$$

we finally get for $(e^{x+y} - e^x - e^y + 1)/xy$ the branched continued fraction

$$\cfrac{1}{1 + \left(\cfrac{-x/2}{1} + \cfrac{x/6}{1} + \cfrac{-x/6}{1} + \dots \right) + \left(\cfrac{-y/2}{1} + \cfrac{y/6}{1} + \cfrac{-y/6}{1} + \dots \right) + \cfrac{xy/4}{1 + \left(\cfrac{x/6}{1} + \cfrac{-x/6}{1} + \dots \right) + \left(\cfrac{y/6}{1} + \cfrac{-y/6}{1} + \dots \right)} + \dots}$$

and for e^{x+y}

$$1 + \left(\cfrac{x}{1} + \cfrac{-x/2}{1} + \cfrac{x/6}{1} + \dots \right) + \left(\cfrac{y}{1} + \cfrac{-y/2}{1} + \cfrac{y/6}{1} + \dots \right) +$$

$$\cfrac{xy}{1 + \left(\cfrac{-x/2}{1} + \cfrac{x/6}{1} + \cfrac{-x/6}{1} + \dots \right) + \left(\cfrac{-y/2}{1} + \cfrac{y/6}{1} + \cfrac{-y/6}{1} + \dots \right) + \cfrac{xy/4}{1 + \left(\cfrac{x/6}{1} + \cfrac{-x/6}{1} + \dots \right) + \left(\cfrac{y/6}{1} + \cfrac{-y/6}{1} + \dots \right)} + \dots}$$

Problems.

(1) Prove that

$$P_{n+1}\, Q_{n-1} - P_{n-1}\, Q_{n+1} = (-1)^{n+1}\, a_1 \ldots a_n . b_{n+1}$$

(2) Prove that

$$P_{n+2}\, Q_{n-2} - P_{n-2}\, Q_{n+2} =$$
$$(-1)^n\, a_1 \ldots a_{n-1}\left[b_n \left(b_{n+1}\, b_{n+2} + a_{n+2} \right) + b_{n+2}\, a_{n+1} \right]$$

(3) Prove formula (1.10b.) for the odd part of a continued fraction.

(4) The convergents of the continued fraction

$$b_0 + \sum_{i=1}^{\infty}\ \frac{1}{|b_i}$$

with $b_i > 0$ for $i \geq 0$ satisfy
  a) $\{C_{2n}\}_{n \in \mathbb{N}}$ is a monotonically increasing sequence.
  b) $\{C_{2n+1}\}_{n \in \mathbb{N}}$ is a monotonically decreasing sequence.
  c) for n and m arbitrary: $C_{2m+1} > C_{2n}$

(5) a) The nth numerator and denominator of the continued fraction

$$b_0 + \sum_{i=1}^{\infty}\ \frac{a_i}{|b_i}$$

satisfy

$$P_n = \begin{vmatrix} b_0 & -1 & 0 & \ldots & 0 \\ a_1 & b_1 & -1 & & \vdots \\ 0 & a_2 & b_2 & \ddots & 0 \\ \vdots & & \ddots & \ddots & -1 \\ 0 & \ldots & 0 & a_n & b_n \end{vmatrix}$$

$$Q_n = \begin{vmatrix} b_1 & -1 & 0 & \ldots & & 0 \\ a_2 & b_2 & -1 & & & \vdots \\ 0 & a_3 & b_3 & \ddots & & 0 \\ \vdots & & & \ddots & \ddots & -1 \\ 0 & \ldots & & 0 & a_n & b_n \end{vmatrix}$$

b) Also

$$Q_n = \frac{\partial P_n}{\partial b_0}$$

c) Prove theorem 1.11.

(6) a) Construct a continued fraction with convergents

$$C_n = (1 + \gamma_0)(1 + \gamma_1)\ldots(1 + \gamma_n)$$

where $\gamma_k(1 + \gamma_k) \neq 0$ for $k \geq 0$.

b) Use it to give a continued fraction expansion for

$$\frac{\sin(\pi x)}{\pi x} = (1 - x)(1 + x)\left(1 - \frac{x}{2}\right)\left(1 + \frac{x}{2}\right)\left(1 - \frac{x}{3}\right)\left(1 + \frac{x}{3}\right)\ldots$$

(7) If

$$C_n = \frac{P_n}{Q_n}$$

are the convergents of a given continued fraction, construct a continued fraction with convergents

$$C_0, \ldots, C_{n-1}, \frac{P_n - \alpha P_{n-1}}{Q_n - \alpha Q_{n-1}}, \quad C_n, C_{n+1}, \ldots$$

where $\alpha \in \mathbf{R}$. This procedure is called the **extension** of a continued fraction.

(8) How is the method of Viscovatov to be adapted if $d_{20} = 0$?

(9) Give a continued fraction representation for $\sqrt{13}$ and discuss its convergence.

(10) Prove formula (1.26.) using (1.24.) $n - 1$ times.

Remarks.

(1)   Facts about the history of continued fractions can be found in [5]. This
        history goes back to Euclids algorithm to compute the greatest common
        divisor of two integers (300 B.C.) but the first conscious use of continued
        fractions dates from the 16th century.

(2)   The notion of nth numerator and denominator satisfying a three-term
        recurrence relation can be generalized to compute solutions of a $(k + 2) -$
        term recurrence relation with $k + 1$ initial data:

$$R_{n+k+1} = b_n R_{n+k} + a_n^{(k)} R_{n+k-1} + \ldots + a_n^{(1)} R_n$$

        The $(k + 1)$-tuple of elements

$$\left(\{a_n^{(1)}\}_n, \ldots, \{a_n^{(k)}\}_n, \{b_n\}_n \right)$$

        is then called a **generalized continued fraction** [26].

(3)   More general forms of continued fractions where a_i and b_i are no longer
        real or complex numbers, are possible. We refer to the works of Fair [8],
        Hayden [9], Roach [19], Wynn [30, 31] and Zemanian [32].
        A lot of references on the theory of continued fractions can also be found
        in the bibliographies edited by Brezinski [4].

(4)   If a continued fraction

$$b_0 + \sum_{i=1}^{\infty} \frac{a_i}{b_i}\Bigg|$$

        with n^{th} convergent C_n, converges to a finite limit C, then $C - C_n$ is
        called the nth **truncation error**. An extensive analysis of truncation
        errors is given in [11 pp. 297-326].

(5)   Another type of bivariate continued fraction expansions can for instance
        be found in [22]. They are of the form

$$B_0(xy) + \sum_{i=1}^{\infty} \frac{a_i^{(1)} x}{\left| B_i^{(1)}(xy) \right.} + \sum_{i=1}^{\infty} \frac{a_i^{(2)} y}{\left| B_i^{(2)}(xy) \right.}$$

where the continued fractions $B_i^{(j)}(xy)$ are given by

$$B_i^{(j)}(xy) = 1 + \sum_{k=1}^{\infty} \frac{b_{i,k}^{(j)} xy}{\left| 1 \right.} \qquad \begin{array}{l} i = 1, 2, \ldots \\ j = 1, 2 \end{array}$$

and obtained by inverting power series. More types of branched continued fractions are given in [3] and [13].

References.

[1] *Abramowitz M.* and *Stegun I.* Handbook of Mathematical functions. Dover publications, New York, 1968.

[2] *Blanch G.* Numerical evaluation of continued fractions. SIAM Rev. 6, 1964, 383-421.

[3] *Bodnarčuk P.* and *Skorobogatko W.* (in Russian) Branched continued fractions and their applications. Naukowaja Dumka, Kiev, 1974.

[4] *Brezinski C.* A bibliography on Padé approximation and related subjects. Publications ANO, Université de Lille, France.

[5] *Brezinski C.* History of continued fractions and Padé approximants. Springer, Heidelberg, 1986.

[6] *Cuyt A.* and *Van der Cruyssen P.* Rounding error analysis for forward continued fraction algorithms. Comput. Math. Appl. 11, 1985, 541-564.

[7] *de Bruin M.* and *van Rossum H.* Padé Approximation and its applications. Lecture Notes in Mathematics 888, Springer , Berlin, 1981.

[8] *Fair W.* Noncommutative continued fractions. SIAM J. Math. Anal. 2, 1971, 226-232.

[9] *Hayden T.* Continued fractions in Banach spaces. Rocky Mountain J. Math. 4, 1974, 357-370.

[10] *Henrici P.* Applied and computational complex analysis: vol. 2. John Wiley, New York, 1976.

[11] *Jones W.* and *Thron W.* Continued fractions: analytic theory and applications. Encyclopedia of Mathematics and its applications: vol. 11, Addison-Wesley, Reading, 1980.

[12] *Khovanskii A.* The application of continued fractions and their generalizations to problems in approximation theory. Noordhoff, Groningen, 1963.

[13] *Kuchminskaya K.* (in Russian) Corresponding and associated branched continued fractions for double power series. Dokl. Akad. Nauk Ukrain.SSR Ser. A 7, 1978, 614-617.

[14] *Mikloško J.*   Investigation of algorithms for numerical computation of continued fractions. USSR Computational Math. and Math. Phys. 16, 1976, 1-12.

[15] *Murphy J.* and *O'Donohoe M.*   A two-variable generalization of the Stieltjes-type continued fraction. J. Comput. Appl. Math. 4, 1978, 181-190.

[16] *Murphy J.* and *O'Donohoe M.*   Some properties of continued fractions with applications in Markov processes. J. Inst. Math. Appl. 16, 1975, 57-71.

[17] *Perron O.*   Die Lehre von den Kettenbruchen II. Teubner, Stuttgart, 1977.

[18] *Pringsheim A.*   Über die Konvergenz unendlicher Kettenbrüche. S.-B.-Bayer. Akad. Wiss. Math.-Nat. Kl. 28, 1899, 295-324.

[19] *Roach F.*   Continued fractions over an inner product space. AMS Proceedings 24, 1970, 576-582.

[20] *Sauer R.* and *Szabó F.*   Mathematische Hilfsmittel des Ingenieurs III. Springer, Berlin, 1968.

[21] *Seidel L.*   Untersuchungen über die Konvergenz und Divergenz der Kettenbrüche. Habilschrift, München, 1846.

[22] *Siemaszko W.*   Branched continued fractions for double power series. J. Comput. Appl. Math. 6, 1980, 121-125.

[23] *Stieltjes T.*   Recherches sur les fractions continues. Ann. Fac. Sci. Toulouse 8, 1894, 1-22 and 9, 1894, 1-47.

[24] *Thron W.* and *Waadeland H.*   Accelerating Convergence of Limit Periodic Continued Fractions K(a_n/1). Numer. Math. 34, 1980, 155-170.

[25] *Van der Cruyssen P.*   A continued fraction algorithm. Numer. Math. 37, 1981, 149-156.

[26] *Van der Cruyssen P.*   Linear Difference Equations and Generalized Continued Fractions. Computing 22, 1979, 269-278.

[27] *Viscovatov B.*   De la méthode générale pour reduire toutes sortes de quantités en fractions continues. Mém. Acad. Impériale Sci. St-Petersburg 1, 1803-1806, 226-247.

[28] *Wall H.*   Analytic theory of continued fractions. Chelsea, Bronx, 1973.

[29] *Worpitzky J.*   Untersuchungen über die Entwickelung der monodronen und monogenen Funktionen durch Kettenbrüche. Friedrichs-Gymnasium und Realschule Jahresbericht, 1865, 3-39.

[30] *Wynn P.*   Continued Fractions whose coefficients obey a non-commutative law of multiplication. Arch. Rational Mech. Anal. 12, 1963, 273-312.

[31] *Wynn P.*   Vector continued fractions. Linear Algebra 1, 1968, 357-395.

[32] *Zemanian A.*   Continued fractions of operator-valued analytic functions. J. Approx. Theory 11, 1974, 319-326.

CHAPTER II: Padé Approximants.

"Dans les applications du calcul, il est ordinairement inutile, sinon impossible, d'obtenir le résultat avec une complète exactitude. Le plus souvent, il suffit seulement d'en connaître une valeur approchée telle que l'erreur commise en adoptant cette valeur au lieu du résultat exact soit inférieure à une limite donnée à priori."

H. PADÉ — *"Sur la réprésentation approchée d'une fonction par des fractions rationelles" (1892).*

§1. Notations and definitions.

Consider a formal power series

$$f(x) = c_0 + c_1 x + c_2 x^2 + \dots \tag{2.1.}$$

with $c_0 \neq 0$.

In the sequel of the text we shall write ∂p for the exact degree of a polynomial p and ωp for the order of a power series p (i.e. the degree of the first nonzero term).

The **Padé approximation problem** of order (m, n) for f consists in finding polynomials

$$p(x) = \sum_{i=0}^{m} a_i x^i$$

and

$$q(x) = \sum_{i=0}^{n} b_i x^i$$

such that in the power series $(f q - p)(x)$ the coefficients of x^i for $i = 0, \dots, m + n$ disappear, i.e.

$$\begin{cases} \partial p \leq m \\ \partial q \leq n \\ \omega(f q - p) \geq m + n + 1 \end{cases} \tag{2.2.}$$

Condition (2.2.) is equivalent with the following two linear systems of equations

$$\begin{cases} c_0 \, b_0 = a_0 \\ c_1 \, b_0 + c_0 \, b_1 = a_1 \\ \quad \vdots \\ c_m \, b_0 + c_{m-1} \, b_1 + \dots + c_{m-n} \, b_n = a_m \end{cases} \tag{2.3a.}$$

$$\begin{cases} c_{m+1} \, b_0 + c_m \, b_1 + \dots + c_{m-n+1} \, b_n = 0 \\ \quad \vdots \\ c_{m+n} \, b_0 + c_{m+n-1} \, b_1 + \dots + c_m \, b_n = 0 \end{cases} \tag{2.3b.}$$

with $c_i = 0$ for $i < 0$.

For $n = 0$ the system of equations (2.3b.) is empty. In this case $a_i = c_i$ $(i = 0, \ldots, m)$ and $b_0 = 1$ satisfy (2.2.), in other words the partial sums of (2.1.) solve the Padé approximation problem of order $(m, 0)$.

In general a solution for the coefficients a_i is known after substitution of a solution for the b_i in the left hand side of (2.3a.). So the crucial point is to solve the homogeneous system of n equations (2.3b.) in the $n + 1$ unknowns b_i. This system has at least one nontrivial solution because one of the unknowns can be chosen freely.

The following relationship can be proved for different solutions of the same Padé approximation problem.

Theorem 2.1.

If the polynomials p_1, q_1 and p_2, q_2 satisfy (2.2.), then $p_1\, q_2 = p_2\, q_1$.

Proof

The polynomial $p_1\, q_2 - p_2\, q_1$ can also be written as

$$(f\, q_2 - p_2)q_1 - (f\, q_1 - p_1)q_2$$

Since

$$\omega(f\, q_1 - p_1) \geq m + n + 1$$
$$\omega(f\, q_2 - p_2) \geq m + n + 1$$

we have

$$\omega(p_1\, q_2 - p_2\, q_1) \geq m + n + 1$$

But $(p_1\, q_2 - p_2 q_1)(x)$ is a polynomial of degree at most $m + n$.
So $p_1\, q_2 = p_2\, q_1$.   ■

A consequence of this theorem is that the rational forms p_1/q_1 and p_2/q_2 are equivalent. Hence all nontrivial solutions of (2.2.) supply the same irreducible form. If $p(x)$ and $q(x)$ satisfy (2.2.) we shall denote by

$$r_{m,n}(x) = \frac{p_0}{q_0}(x)$$

the irreducible form of p/q normalized such that $q_0(0) = 1$.

This rational function $r_{m,n}(x)$ is called the **Padé approximant** of order (m, n) for f. By calculating the irreducible form, a polynomial may be cancelled in numerator and denominator of p/q. We shall therefore denote the exact degrees of p_0 and q_0 in $r_{m,n}$ respectively by $\mathbf{m'}$ and $\mathbf{n'}$.

As a conclusion we can formulate the next theorem.

Theorem 2.2.

For every nonnegative m and n a unique Padé approximant of order (m, n) for f exists.

§2. Fundamental properties.

2.1. Properties of the Padé approximant.

Let $r_{m,n} = p_0/q_0$ be the Padé approximant of order (m, n) for f. Although p_0 and q_0 are computed from polynomials p and q that satisfy (2.2.), it is not necessary that p_0 and q_0 satisfy (2.2.) themselves. A simple example will illustrate this.

Consider $f(x) = 1 + x^2$ and take $m = 1 = n$. Condition (2.2.) is then equivalent with

$$\begin{cases} b_0 = a_0 \\ b_1 = a_1 \end{cases}$$

$$\{ b_0 = 0$$

A solution is given by $b_0 = 0 = a_0$ and $b_1 = 1 = a_1$. So $p(x) = x = q(x)$. Consequently $p_0 = 1 = q_0$ with $\omega(f\, q_0 - p_0) = 2 < m + n + 1$ and the corresponding equations (2.2.) do not hold.

But it is easy to construct, from the knowledge of p_0 and q_0, a solution of the system of equations (2.2.).

Theorem 2.3.

If the Padé approximant of order (m, n) for f is given by

$$r_{m,n}(x) = \frac{p_0}{q_0}(x)$$

then there exists an integer s with $0 \le s \le \min(m - m', n - n')$, such that $p(x) = x^s\, p_0(x)$ and $q(x) = x^s\, q_0(x)$ satisfy (2.2.).

Proof

Let p_1, q_1 be a nontrivial solution of (2.2.).
Hence

$$\partial p_1 \le m$$
$$\partial q_1 \le n$$
$$\omega(f\, q_1 - p_1) \ge m + n + 1$$

Since the irreducible form of p_1/q_1 is p_0/q_0 we know that

$$p_1(x) = t(x)\, p_0(x)$$
$$q_1(x) = t(x)\, q_0(x)$$

with $t(x)$ a polynomial of degree at most $\min(m - m', n - n')$. If s is the order of the polynomial $t(x)$, then

$$0 \le s \le \min(m - m', n - n')$$

Since

$$\begin{aligned}
\omega(f\, q_1 - p_1) &= \omega[t(f\, q_0 - p_0)] \\
&= \omega[x^s(f\, q_0 - p_0)] \\
&= \omega[f(x^s\, q_0) - (x^s\, p_0)]
\end{aligned}$$

the proof is completed. ∎

For another definition of Padé approximant we refer to problem (2) at the end of this chapter. We have defined here the Padé approximation problem for a formal power series given around the origin. Of course it is possible to consider the same problem in any other point. Let

$$f(x) = \sum_{i=0}^{\infty} c_i\, (x - x_0)^i$$

be a formal power series around the point x_0 in the complex plane, with $c_0 \ne 0$. It is possible to construct nontrivial polynomials

$$p(x) = \sum_{i=0}^{m} a_i(x - x_0)^i$$

and

$$q(x) = \sum_{i=0}^{n} b_i(x - x_0)^i$$

such that

$$(f\, q - p)(x) = \sum_{i \ge m+n+1} d_i(x - x_0)^i$$

This condition results in the same two systems of equations (2.3a.) and (2.3b.) and similar results hold.

2.2. Block structure of the Padé table.

The Padé approximants $r_{m,n}$ for f can be ordered in a table for different values of m and n:

$$
\begin{array}{ccccc}
r_{0,0} & r_{0,1} & r_{0,2} & r_{0,3} & \cdots \\[2mm]
r_{1,0} & r_{1,1} & r_{1,2} & r_{1,3} & \cdots \\[2mm]
r_{2,0} & r_{2,1} & r_{2,2} & r_{2,3} & \cdots \\[2mm]
r_{3,0} & r_{3,1} & r_{3,2} & & \cdots \\[2mm]
r_{4,0} & r_{4,1} & & \cdots \\[2mm]
r_{5,0} \\[2mm]
\vdots
\end{array}
$$

This table is called the **Padé table** of f. As we already remarked, the first column consists of the partial sums of f. The first row contains the reciprocals of the partial sums of $1/f$ (see also problem (4)).
We now give part of the Padé table for $\exp(x)$ (table 2.1.) and for $1 + \sin(x)$ (table 2.2.). In table 2.1. all the Padé approximants are different while in table 2.2. certain $r_{m,n}$ coincide. In general the following result can be proved.

Theorem 2.4.

Let the Padé approximant of order (m, n) for f be given by

$$
r_{m,n}(x) = \frac{p_0}{q_0}(x)
$$

Then

(a) $\omega(f\, q_0 - p_0) = m' + n' + t + 1$ with $t \geq 0$

(b)   for k and ℓ satisfying $m' \leq k \leq m' + t$ and $n' \leq \ell \leq n' + t$:

$$
r_{k,\ell}(x) = \frac{p_0}{q_0}(x)
$$

(c) $m \leq m' + t$ and $n \leq n' + t$

Table 2.1.

$$f(x) = e^x = 1 + x + \frac{x^2}{2!} + \frac{x^3}{3!} + \frac{x^4}{4!} + \ldots$$

1	$\dfrac{1}{1-x}$	$\dfrac{1}{1-x+\frac{1}{2}x^2}$	$\ldots$
$1+x$	$\dfrac{1+\frac{1}{2}x}{1-\frac{1}{2}x}$	$\dfrac{1+\frac{1}{3}x}{1-\frac{2}{3}x+\frac{1}{6}x^2}$	$\ldots$
$1+x+\frac{1}{2}x^2$	$\dfrac{1+\frac{2}{3}x+\frac{1}{6}x^2}{1-\frac{1}{3}x}$	$\dfrac{1+\frac{1}{2}x+\frac{1}{12}x^2}{1-\frac{1}{2}x+\frac{1}{12}x^2}$	$\ldots$
$1+x+\frac{1}{2}x^2+\frac{1}{6}x^3$	$\dfrac{1+\frac{3}{4}x+\frac{1}{4}x^2+\frac{1}{24}x^3}{1-\frac{1}{4}x}$	$\vdots$	
$1+x+\frac{1}{2}x^2+\frac{1}{6}x^3+\frac{1}{24}x^4$	$\vdots$		
$\vdots$			

Table 2.2.

$$f(x) = 1 + \sin(x) = 1 + x - \frac{x^3}{3!} + \frac{x^5}{5!} - \frac{x^7}{7!} + \ldots$$

$$1 \qquad\qquad \frac{1}{1-x} \qquad\qquad \frac{1}{1-x+x^2} \qquad\qquad \frac{1}{1-x+x^2-\frac{5}{6}x^3} \qquad \ldots$$

$$1+x \qquad\qquad 1+x \qquad\qquad \frac{1+\frac{5}{6}x}{1-\frac{1}{6}x+\frac{1}{6}x^2} \qquad\qquad \ldots$$

$$1+x \qquad\qquad 1+x \qquad\qquad \frac{1+x+\frac{1}{6}x^2}{1+\frac{1}{6}x^2} \qquad\qquad \ldots$$

$$1+x-\tfrac{1}{6}x^3 \qquad\qquad 1+x-\tfrac{1}{6}x^3 \qquad\qquad \vdots$$

$$1+x-\tfrac{1}{6}x^3 \qquad\qquad 1+x-\tfrac{1}{6}x^3$$

$$1+x-\tfrac{1}{6}x^3+\tfrac{1}{120}x^5 \qquad\qquad \vdots$$

$$\vdots$$

Proof

To prove (a) we use theorem 2.3.:

$$\omega(f\, q_0 - p_0) \geq m + n + 1 - s \quad \text{with} \quad 0 \leq s \leq \min(m - m', n - n')$$

This implies

$$\omega(f\, q_0 - p_0) \geq m' + n' + 1$$

or

$$\omega(f\, q_0 - p_0) = m' + n' + 1 + t$$

with $t \geq 0$ and t as large as possible.
Let us now consider integers k and ℓ such that

$$m' \leq k \leq m' + t$$
$$n' \leq \ell \leq n' + t$$

We put

$$s = \min(k - m', \ell - n')$$
$$p(x) = p_0(x)\, x^s$$
$$q(x) = q_0(x)\, x^s$$

Then clearly

$$\partial p \leq k$$
$$\partial q \leq \ell$$

and as a consequence of (a),

$$\omega(f\, q - p) = m' + n' + t + s + 1$$

Since

$$m' + n' + t + s + 1 \geq k + \ell + 1$$

we have

$$r_{k,\ell}(x) = \frac{p_0}{q_0}(x)$$

because p, q is a solution of the Padé approximation problem of order (k, ℓ). To prove (c) we again use theorem 2.3. It guarantees the existence of an integer s with $0 \leq s \leq \min(m - m', n - n')$ such that

$$m + n + 1 \leq \omega(f\,q - p) = m' + n' + t + s + 1$$

Since $s \leq m - m'$ we have

$$m + n + 1 \leq m + n' + t + 1$$

or

$$n \leq n' + t$$

Analogously

$$m \leq m' + t \qquad\qquad \blacksquare$$

The previous property is called the **block structure** of the Padé table: the table consists of square blocks of size $(t + 1)$ containing equal Padé approximants.

2.3. Normality.

As a result of theorem 2.4. we call a Padé approximant **normal** if it occurs only once in the Padé table. A criterion for the normality of an approximant is given in the next theorem.

Theorem 2.5.

The Padé approximant

$$r_{m,n} = \frac{p_0}{q_0}$$

for f is normal if and only if

(a) $m = m'$ and $n = n'$

(b) $\omega(f\,q_0 - p_0) = m + n + 1$

Proof

If $r_{m,n}$ is normal then (a) and (b) can be proved by contraposition. Suppose $m' < m$ or $n' < n$. Then by theorem 2.4.(a):

$$\omega(f\, q_0 - p_0) \geq m' + n' + 1$$

This implies

$$r_{m',n'}(x) = \frac{p_0}{q_0}(x)$$

which is a contradiction with the normality of $r_{m,n}(x)$.
Suppose (b) is not valid. Using theorem 2.4. (a) we know that

$$\omega(f\, q_0 - p_0) = m' + n' + 1 + t$$

with $t > 0$.
Since (a) must be valid:

$$\omega(f\, q_0 - p_0) = m + n + 1 + t$$

with $t > 0$.
Hence

$$r_{k,\ell}(x) = \frac{p_0}{q_0}(x)$$

for all k and ℓ satisfying $m' \leq k \leq m' + t$ and $n' \leq \ell \leq n' + t$.
This again contradicts the normality of $r_{m,n}(x)$.
To prove that (a) and (b) guarantee the normality of $r_{m,n}(x)$ we proceed as follows. Suppose $r_{m,n}(x) = r_{k,\ell}(x)$ for certain k and ℓ with $k > m$ or $\ell > n$. For an integer s that satisfies

$$\omega[(f\, q_0 - p_0)\, x^s] \geq k + \ell + 1$$

we find, by using (b), that

$$s > k - m \quad \text{or} \quad s > \ell - n$$

This contradicts theorem 2.3. ∎

Normality of a Padé approximant can also be guaranteed by the nonvanishing of certain determinants.

We introduce the notation

$$D_{m,n+1} = \begin{pmatrix} c_m & c_{m-1} & \cdots & c_{m-n} \\ c_{m+1} & c_m & \cdots & c_{m+1-n} \\ \vdots & \vdots & \ddots & \vdots \\ c_{m+n} & c_{m+n-1} & \cdots & c_m \end{pmatrix}$$

with $\det D_{m,0} = 1$. The following result can be proved [40 p. 243].

Theorem 2.6.

The Padé approximant

$$r_{m,n} = \frac{p_0}{q_0}$$

for f is normal if and only if

$$\det\ D_{m,n} \neq 0$$
$$\det\ D_{m+1,n} \neq 0$$
$$\det\ D_{m,n+1} \neq 0$$
$$\det\ D_{m+1,n+1} \neq 0$$

For Stieltjes series this theorem and the following lemma [30 p. 605] lead to a remarkable result.

Lemma 2.1.

Let $g(t)$ be a real-valued, bounded, nondecreasing function defined on $[0, \infty)$ and let the integrals

$$c_i = \int_0^\infty t^i dg(t)$$

exist for all $i \geq 0$. If $g(t)$ has at least k points of increase then for all $m \geq 0$ and for $n = 0, \ldots, k$ we have

$$\det D_{m,n} \neq 0$$

If $g(t)$ has an infinite number of points of increase then for all $m, n \geq 0$

$$\det D_{m,n} \neq 0$$

Clearly for Stieltjes series

$$\sum_{i=0}^{\infty} \left(\int_{0}^{\infty} t^i dg(t) \right) x^i$$

with $g(t)$ having infinitely many points of increase, the latter is true and hence we can conclude the following.

Theorem 2.7.

Let f be a Stieltjes series and let g be a real-valued, bounded, nondecreasing function having infinitely many points of increase. Then for all $m, n \geq 0$ the Padé approximant $r_{m,n}$ for

$$f(x) = \sum_{i=0}^{\infty} \left(\int_{0}^{\infty} t^i dg(t) \right) x^i$$

is normal.

§3. Methods to compute Padé approximants.

In the sequel of the text we suppose that every Padé approximant in the Padé table itself satisfies the condition (2.2.). By theorem 2.3. this is the case if for instance $\min(m - m', n - n') = 0$ for all m and n.
A survey of algorithms for computing Padé approximants is given in [50] and [11].

3.1. Corresponding continued fractions.

The following theorem shall be used to compute the difference of neighbouring Padé approximants in the table.

Theorem 2.8.

If

$$r_{m,n} = \frac{p_1}{q_1}$$

and

$$r_{m+k,n+\ell} = \frac{p_2}{q_2}$$

with $k, \ell \geq 0$ then a polynomial $v(x)$ exists with

$$\partial v \leq \max(k - 1, \ell - 1)$$

$$(p_1 q_2 - p_2 q_1)(x) = x^{m+n+1}\, v(x)$$

Proof

For the expression $p_1 q_2 - p_2 q_1$ we can write

$$
\begin{aligned}
\omega(p_1 q_2 - p_2 q_1) &= \omega[(fq_2 - p_2)q_1 - (fq_1 - p_1)q_2] \\
&\geq \min(\omega(fq_2 - p_2), \omega(fq_1 - p_1)) \\
&\geq m + n + 1 \\
\partial(p_1 q_2 - p_2 q_1) &\leq \max(\partial p_1 + \partial q_2, \partial p_2 + \partial q_1) \\
&\leq m + n + \max(k, \ell)
\end{aligned}
$$

This completes the proof.   ∎

Let us now consider the following sequence of elements on a descending staircase in the Padé table

$$T_k = \{r_{k,0}, r_{k+1,0}, r_{k+1,1}, r_{k+2,1}, \ldots\} \quad \text{for} \quad k \geq 0$$

and the following continued fraction

$$d_0 + d_1 x + \ldots + d_k x^k + \left|\frac{d_{k+1} x^{k+1}}{1}\right. + \left|\frac{d_{k+2} x}{1}\right. + \left|\frac{d_{k+3} x}{1}\right. + \ldots \qquad (2.4.)$$

Theorem 2.9.

If every three consecutive elements in T_k are different, then a continued fraction of the form (2.4.) exists with $d_{k+i} \neq 0$ for $i \geq 1$ and such that the nth convergent equals the $(n+1)^{th}$ element of T_k.

Proof

Put

$$r_{k+i,j} = \frac{P_{i+j}}{Q_{i+j}}$$

for $i = j, j+1$ and $j = 0, 1, 2, \ldots$.
A continued fraction of which the nth convergent equals

$$\frac{P_n}{Q_n}(n \geq 0)$$

is according to theorem 1.4., given by

$$P_0 + \left|\frac{P_1 - P_0}{1}\right. + \sum_{i=1}^{\infty} \left|\frac{\dfrac{P_i Q_{i+1} - P_{i+1} Q_i}{P_i Q_{i-1} - P_{i-1} Q_i}}{\dfrac{P_{i+1} Q_{i-1} - P_{i-1} Q_{i+1}}{P_i Q_{i-1} - P_{i-1} Q_i}}\right. \qquad (2.5a.)$$

Here we have already used the fact that $Q_0 = Q_1 = 1$. By theorem 2.8. we find that

$$\frac{P_i Q_{i+1} - P_{i+1} Q_i}{P_i Q_{i-1} - P_{i-1} Q_i} = a_i x$$

$$i = 1, 2, \ldots$$

$$\frac{P_{i+1} Q_{i-1} - P_{i-1} Q_{i+1}}{P_i Q_{i-1} - P_{i-1} Q_i} = b_i$$

for certain nonzero numbers a_i and b_i.
So the continued fraction (2.5a.) is

$$P_0 + \left| \frac{P_1 - P_0}{1} \right| + \sum_{i=1}^{\infty} \left| \frac{a_i x}{b_i} \right|$$

with

$$P_0 = r_{k,0} = \sum_{i=0}^{k} c_i x^i$$

and

$$P_1 = r_{k+1,0} = \sum_{i=0}^{k+1} c_i x^i$$

By performing an equivalence transformation we finally get a continued fraction of the form (2.4.). ∎

In this way we are able to construct corresponding continued fractions for functions f analytic in the origin.

Theorem 2.10.

If the nth convergent of (2.4.) equals the $(n+1)^{th}$ element of T_0 $(n \geq 0)$, then (2.4.) is the corresponding continued fraction to the power series (2.2.).

Proof

Let P_n/Q_n be the nth convergent of (2.4.).
Then

$$\omega(f Q_n - P_n) \geq n + 1 \text{ and } Q_n(0) = 1$$

because P_n/Q_n is also the $(n+1)^{th}$ element of T_0.

Hence

$$\left(f - \frac{P_n}{Q_n}\right)^{(j)}(0) = 0 \quad j = 0, \ldots, n$$

because $Q_n(x)$ is nontrivial in a neighbourhood of the origin. So the Taylor series development of the nth convergent matches the given power series up to and including the term of degree n. In other words, (2.4.) is the corresponding continued fraction to (2.1.). ∎

By continued fractions of the form (2.4.) one can only compute Padé approximants below the main diagonal in the Padé table. For the right upper half of the table one can use the reciprocal covariance property of Padé approximants, given in problem (4) at the end of this chapter.

We now turn to the problem of the calculation of the coefficients d_{k+i} in (2.4.) for $i > 1$ starting from the coefficients $c_0, c_1, c_2, \ldots$, and not from the knowledge of the Padé approximants.

3.2. The qd-algorithm.

Consider the following continued fraction

$$g_k(x) = c_0 + \ldots + c_k\,x^k + \left.\frac{c_{k+1}x^{k+1}}{1}\right| - \left.\frac{q_1^{(k+1)}x}{1}\right| - \left.\frac{e_1^{(k+1)}x}{1}\right| -$$

$$\left.\frac{q_2^{(k+1)}x}{1}\right| - \left.\frac{e_2^{(k+1)}x}{1}\right| - \ldots \quad (2.5b.)$$

If the coefficients $q_\ell^{(k+1)}$ and $e_\ell^{(k+1)}$ are computed as in theorem 2.9. then the convergents of g_k equal the elements of T_k.

If we calculate the even part of $g_k(x)$ we get

$$c_0 + \ldots + c_k x^k + \left.\frac{c_{k+1}x^{k+1}}{1 - q_1^{(k+1)}x}\right| + \left.\frac{-q_1^{(k+1)}e_1^{(k+1)}x^2}{1 - (q_2^{(k+1)} + e_1^{(k+1)})x}\right| + \left.\frac{-q_2^{(k+1)}e_2^{(k+1)}x^2}{1 - (q_3^{(k+1)} + e_2^{(k+1)})x}\right| + \ldots$$

If we calculate the odd part of $g_{k-1}(x)$ we get

$$c_0 + \ldots + c_k\, x^k + \cfrac{c_k\, q_1^{(k)} x^{k+1}}{1 - (q_1^{(k)} + e_1^{(k)})x} + \cfrac{-e_1^{(k)} q_2^{(k)} x^2}{1 - (q_2^{(k)} + e_2^{(k)})x} + \cfrac{-e_2^{(k)} q_3^{(k)} x^2}{1 - (q_3^{(k)} + e_3^{(k)})x} + \ldots$$

The even part of $g_k(x)$ and the odd part of $g_{k-1}(x)$ are two continued fractions which have the same convergents $r_{k,0}, r_{k+1,1}, r_{k+2,2}, \ldots$ and which also have the same form. Hence the partial numerators and denominators must be equal, and we obtain [43] for $k \geq 1$ and $\ell \geq 1$

$$q_1^{(k)} = \frac{c_{k+1}}{c_k}$$

$$e_\ell^{(k)} = e_{\ell-1}^{(k+1)} + q_\ell^{(k+1)} - q_\ell^{(k)} \tag{2.6a.}$$

$$q_{\ell+1}^{(k)} = q_\ell^{(k+1)} \frac{e_\ell^{(k+1)}}{e_\ell^{(k)}} \tag{2.6b.}$$

with

$$e_0^{(k)} = 0$$

The numbers $q_\ell^{(k)}$ and $e_\ell^{(k)}$ are usually arranged in a table, where the superscript (k) indicates a diagonal and the subscript ℓ indicates a column. This table is called the qd-table.

Table 2.3.

$e_0^{(1)}$					
	$q_1^{(1)}$				
$e_0^{(2)}$		$e_1^{(1)}$			
	$q_1^{(2)}$		$q_2^{(1)}$		
$e_0^{(3)}$		$e_1^{(2)}$		$e_2^{(1)}$	
	$q_1^{(3)}$		$q_2^{(2)}$		$\ddots$
$e_0^{(4)}$	$\vdots$	$e_1^{(3)}$	$\vdots$	$e_2^{(2)}$	
$\vdots$		$\vdots$		$\vdots$	

The formulas (2.6.) can now also be memorized as follows: $e_\ell^{(k)}$ is calculated such that in the following rhombus the sum of the two elements on the upper diagonal equals the sum of the two elements on the lower diagonal

$$q_\ell^{(k)}$$
$$+$$
$$e_{\ell-1}^{(k+1)} \qquad\qquad\qquad e_\ell^{(k)}$$
$$+$$
$$q_\ell^{(k+1)}$$

and $q_{\ell+1}^{(k)}$ is computed such that in the next rhombus the product of the two elements on the upper diagonal equals the product of the two elements on the lower diagonal

$$e_\ell^{(k)}$$
$$*$$
$$q_\ell^{(k+1)} \qquad\qquad\qquad q_{\ell+1}^{(k)}$$
$$*$$
$$e_\ell^{(k+1)}$$

Since the qd-algorithm computes the coefficients in (2.4.), it can be used to compute the Padé approximants below the main diagonal in the Padé table. To calculate the Padé approximants in the right upper half of the table, the qd-algorithm itself can be extended above the diagonal and the following results can be proved [32 pp. 615-617].

Table 2.4.

	$q_1^{(0)}$		$q_2^{(-1)}$		$q_3^{(-2)}$	$\cdots$
$e_0^{(1)}$		$e_1^{(0)}$		$e_2^{(-1)}$		
	$q_1^{(1)}$		$q_2^{(0)}$		$q_3^{(-1)}$	$\cdots$
$e_0^{(2)}$		$e_1^{(1)}$		$e_2^{(0)}$		
	$q_1^{(2)}$		$q_2^{(1)}$		$q_3^{(0)}$	$\cdots$
$e_0^{(3)}$		$e_1^{(2)}$		$e_2^{(1)}$		
	$q_1^{(3)}$		$q_2^{(2)}$		$q_3^{(1)}$	$\cdots$
$e_0^{(4)}$	$\vdots$	$e_1^{(3)}$	$\vdots$	$e_2^{(2)}$	$\vdots$	
$\vdots$		$\vdots$		$\vdots$		

Let

$$\frac{1}{f}(x) = w_0 + w_1 x + w_2 x^2 + \ldots$$

and put

$$q_1^{(0)} = \frac{-w_1}{w_0}$$

$$e_1^{(0)} = \frac{w_2}{w_1}$$

and for $k \geq 1$

$$q_{k+1}^{(-k)} = 0$$

$$e_{k+1}^{(-k)} = \frac{w_{k+2}}{w_{k+1}}$$

If the elements in the extended qd-table are all calculated by the use of (2.6.) using the above starting values, then the continued fraction

$$h_k(x) = \left. \frac{1}{w_0 + w_1 x + \ldots + w_k x^k} \right| + \left. \frac{w_{k+1} x^{k+1}}{1} \right| - \left. \frac{e_{k+1}^{(-k)} x}{1} \right| - \left. \frac{q_{k+2}^{(-k)} x}{1} \right|$$

$$- \left. \frac{e_{k+2}^{(-k)} x}{1} \right| - \left. \frac{q_{k+3}^{(-k)} x}{1} \right| - \ldots$$

supplies the Padé approximants on the staircase

$$U_k = \{r_{0,k}, r_{0,k+1}, r_{1,k+1}, r_{1,k+2}, r_{2,k+2}, \ldots\}$$

To illustrate the qd-scheme we will again calculate some Padé approximants for the function $\exp(x)$. Table 2.3. looks like

$$
\begin{array}{cccccc}
0 & & & & & \\
 & \frac{1}{2} & & & & \\
0 & & -\frac{1}{6} & & & \\
 & \frac{1}{3} & & \frac{1}{6} & & \\
0 & & -\frac{1}{12} & & -\frac{1}{10} & \\
 & \frac{1}{4} & & \frac{3}{20} & & \cdots \\
0 & & -\frac{1}{20} & & \cdots & \\
 & \frac{1}{5} & & \cdots & & \\
0 & & \cdots & & &
\end{array}
$$

From this we get

$$
g_0(x) = 1 + \frac{x}{|1|} - \frac{\frac{1}{2}x}{|1|} + \frac{\frac{1}{6}x}{|1|} - \frac{\frac{1}{6}x}{|1|} + \frac{\frac{1}{10}x}{|1|} + \dots
$$

$$
= 1 + \frac{x}{|1|} + \frac{x}{|-2|} + \frac{x}{|-3|} + \frac{x}{|2|} + \frac{x}{|5|} + \dots
$$

It is obvious that difficulties can arise if the division in (2.6b.) cannot be performed by the fact that $e_\ell^{(k)} = 0$. This is the case if the Padé table is not normal for consecutive elements in T_k can then be equal. Reformulations of the qd-algorithm in this case are given in [13] and [30].

3.3. The algorithm of Gragg.

Let us now consider ascending staircases in the Padé table.
Take

$$
S_k = \{r_{k,0}, \; r_{k-1,0}, \; r_{k-1,1}, \; r_{k-2,1}, \dots, r_{0,k-1}, \; r_{0,k}\}
$$

and consider the continued fraction

$$
f_k(x) = c_0 + \dots + c_k\,x^k - \frac{c_k\,x^k}{|1|} - \frac{f_1^{(k)}}{|x|} - \frac{s_1^{(k)}}{|1|} - \dots - \frac{f_k^{(k)}}{|x|}
$$

In an analogous way as for T_k one can compute the coefficients $f_\ell^{(k)}$ and $s_\ell^{(k)}$ such that the nth convergent of this continued fraction equals the $(n+1)^{th}$ element of S_k. Remark that $f_k(x)$ is not an infinite expression since S_k is a finite sequence. If we compute the odd part of f_{k+1} and the even part of f_k, we again get continued fractions of the same form that have the same convergents. This reasoning provides us with formulas for $f_\ell^{(k)}$ and $s_\ell^{(k)}$ [29]:
for $k \geq 1$

$$s_0^{(k-1)} = 0$$

$$f_k^{(k-1)} = 0$$

$$f_1^{(k)} = \frac{c_{k-1}}{c_k}$$

$$s_k^{(k)} = \frac{w_{k-1}}{w_k}$$

where

$$\frac{1}{f}(x) = w_0 + w_1 x + w_2 x^2 + \ldots$$

and for $k \geq 1$ and $1 \leq \ell \leq k-1$

$$s_\ell^{(k)} = s_{\ell-1}^{(k-1)} + f_\ell^{(k-1)} - f_\ell^{(k)}$$

$$f_{\ell+1}^{(k)} = f_\ell^{(k-1)} \frac{s_\ell^{(k-1)}}{s_\ell^{(k)}}$$

These quantities are arranged in a table as follows. The superscript denotes now an upward sloping diagonal.

Table 2.5.

	0		0		0	$\ldots$
0		$s_1^{(1)}$		$s_2^{(2)}$		
	$f_1^{(1)}$		$f_2^{(2)}$		$f_3^{(3)}$	$\ldots$
0		$s_1^{(2)}$		$s_2^{(3)}$		$\vdots$
	$f_1^{(2)}$		$f_2^{(3)}$		$\vdots$	
0	$\vdots$	$s_1^{(3)}$	$\vdots$			
$\vdots$		$\vdots$				

For the computation of $s_\ell^{(k)}$ and $f_\ell^{(k)}$ we have similar rhombus rules as for the $q_\ell^{(k)}$ and $e_\ell^{(k)}$:

$$f_\ell^{(k)}$$

$$+$$

$$s_{\ell-1}^{(k)} \qquad\qquad s_\ell^{(k+1)}$$

$$+$$

$$f_\ell^{(k+1)}$$

and

$$s_\ell^{(k)}$$

$$*$$

$$f_\ell^{(k)} \qquad\qquad f_{\ell+1}^{(k+1)}$$

$$*$$

$$s_\ell^{(k+1)}$$

3.4. Determinant formulas.

One can also solve the system of equations (2.3b.) and thus get explicit formulas for the Padé approximant.
For

$$f(x) = \sum_{i=0}^{\infty} c_i\, x^i$$

we write

$$F_k(x) = \sum_{i=0}^{k} c_i\, x^i \qquad k \geq 0$$

$$F_k(x) = 0 \qquad\qquad k < 0$$

Theorem 2.11.

If the Padé approximant of order (m, n) for f is given by

$$r_{m,n}(x) = \frac{p_0}{q_0}(x)$$

and if $D = \det\ D_{m,n} \neq 0$, then

$$p_0(x) = \frac{1}{D} \begin{vmatrix} F_m(x) & xF_{m-1}(x) & \cdots & x^n F_{m-n}(x) \\ c_{m+1} & & & \\ \vdots & & D_{m,n} & \\ c_{m+n} & & & \end{vmatrix}$$

and

$$q_0(x) = \frac{1}{D} \begin{vmatrix} 1 & x & \cdots & x^n \\ c_{m+1} & & & \\ \vdots & & D_{m,n} & \\ c_{m+n} & & & \end{vmatrix}$$

Proof

Since $D \neq 0$ the homogeneous system (2.3b.) has a unique solution for the choice $b_0 = 1$. Thus the following homogeneous system has a nontrivial solution:

$$\begin{cases} (1 - q_0(x))b_0 + xb_1 + x^2 b_2 + \cdots + x^n b_n = 0 \\ c_{m+1} b_0 + c_m b_1 + \cdots + c_{m+1-n} b_n = 0 \\ \vdots \\ c_{m+n} b_0 + c_{m+n-1} b_1 + \cdots + c_m b_n = 0 \end{cases}$$

This implies that the determinant of the coefficient matrix of this system is zero:

$$\begin{vmatrix} 1 - q_0(x) & x & \cdots & x^n \\ c_{m+1} & c_m & \cdots & c_{m+1-n} \\ \vdots & \vdots & \ddots & \vdots \\ c_{m+n} & c_{m+n-1} & \cdots & c_m \end{vmatrix} = 0$$

and it proves the formula for $q_0(x)$. If we take a look at $f(x) \, q_0(x)$ we have

$$f(x) \, q_0(x) = \frac{1}{D} \begin{vmatrix} f(x) & xf(x) & \cdots & x^n f(x) \\ c_{m+1} & c_m & \cdots & c_{m+1-n} \\ \vdots & \vdots & \ddots & \vdots \\ c_{m+n} & c_{m+n-1} & \cdots & c_m \end{vmatrix}$$

Because the polynomial $p_0(x)$ contains all the terms of degree less than or equal to m of the series $f(x) \, q_0(x)$, we get the determinant expression for $p_0(x)$ given above. ∎

The determinant formula for $q_0(x)$ can also very easily be obtained by solving (2.3b.) using Cramer's rule after choosing $b_0 = D_{m,n}$. The determinant expressions are of course only useful for small values of m and n because the calculation of a determinant involves a lot of additions and multiplications. They merely exhibit closed form formulas for the solution.

From the proof of theorem 2.11. we can also deduce that

$$(f \, q_0 - p_0)(x) = \frac{1}{D} \begin{vmatrix} \overline{F}_{m+n}(x) & x\overline{F}_{m+n-1}(x) & \cdots & x^n \overline{F}_m(x) \\ c_{m+1} & & & \\ \vdots & & & D_{m,n} \\ c_{m+n} & & & \end{vmatrix}$$

where

$$\overline{F}_k(x) = \sum_{i=k+1}^{\infty} c_i \, x^i = f(x) - F_k(x)$$

This gives an explicit formula for the error $(f - r_{m,n})(x)$ in terms of the coefficients c_i in $f(x)$.

For Stieltjes series

$$\sum_{i=0}^{\infty} \left(\int_0^r t^i dg(t) \right) x^i$$

with convergence radius $\frac{1}{r}$, it is possible using the error formula, to indicate

within a finite set of Padé approximants which one is the most accurate on
$(-\infty, \frac{1}{r}[$. The most interesting result is the following one.

Theorem 2.12.

Let f be a Stieltjes series. For the Padé approximants in the set

$$\{r_{m,n} \mid 0 \leq m + n \leq 2k\}$$

we have

$$\forall x \in (-\infty, \frac{1}{r}[: \ |f(x) - r_{k,k}(x)| = \min_{0 \leq m+n \leq 2k} |f(x) - r_{m,n}(x)|$$

and for those in

$$\{r_{m,n} \mid 0 \leq m + n \leq 2k + 1\}$$

we have

$$\forall x \in (-\infty, \frac{1}{r}[: \ |f(x) - r_{k+1,k}(x)| = \min_{0 \leq m+n \leq 2k+1} |f(x) - r_{m,n}(x)|$$

This means that when k increases, the best Padé approximants for a Stieltjes
series among the elements in successive triangles

$$
\begin{array}{cccc}
r_{0,0} & r_{0,1} & \cdots & r_{0,k} \\
r_{1,0} & & & \\
\vdots & & \cdot\cdot\cdot & \\
r_{k,0} & & &
\end{array}
$$

are the elements on the descending staircase T_0, in other words they are
the successive convergents of the corresponding continued fraction for the
Stieltjes series.

For other subsets of the Padé table similar results exist because, when f is a
Stieltjes series, the errors $(f - r_{m,n})(x)$ are linked by inequalities throughout
the entire table. The interested reader is referred to [8].

3.5. The method of Viscovatov.

By the method of Viscovatov described in section 3.3. of chapter I, the recursive generation of staircase sequences T_k of Padé approximants is absolutely straightforward in the case of a normal Padé table. We have proved in section 3.4. of chapter I that the constructed continued fraction is corresponding. Hence it generates the elements on T_0. If the method of Viscovatov is applied to

$$(f(x) - \sum_{i=0}^{k} c_i x^i)/x^{k+1}$$

for the construction of a corresponding continued fraction

$$\frac{d_{10}}{\left| d_{00} \right.} + \frac{d_{20} x}{\left| d_{10} \right.} + \frac{d_{30} x}{\left| d_{20} \right.} + \ldots$$

then the elements on T_k are obtained from

$$f(x) = \sum_{i=0}^{k} c_i x^i + \frac{d_{10} x^{k+1}}{\left| d_{00} \right.} + \frac{d_{20} x}{\left| d_{10} \right.} + \frac{d_{30} x}{\left| d_{20} \right.} + \ldots$$

However in order to obtain the normalized

$$r_{m,n} = \frac{p_0}{q_0}$$

the normalization $q_0(0) = 1$ has to be built into the algorithm via an equivalence transformation. We reformulate it as follows.
For

$$f(x) = \sum_{i=0}^{\infty} c_{i+k+1} \, x^i$$

we put

$$
\begin{aligned}
d_{1i} &= c_{k+1+i} & i &= 0, 1, 2, \ldots \\
d_{00} &= 1 \\
d_{0i} &= 0 & i &= 1, 2, \ldots
\end{aligned}
$$

and for $j \geq 2$

$$d_{j,i} = \frac{d_{j-2,i+1}}{d_{j-2,0}} - \frac{d_{j-1,i+1}}{d_{j-1,0}}$$

Then

$$f(x) = \sum_{i=0}^{k} c_i \; x^i + \left|\frac{d_{10}x^{k+1}}{1}\right| + \left|\frac{d_{20}x}{1}\right| + \left|\frac{d_{30}x}{1}\right| + \ldots \qquad (2.7.)$$

3.6. Recursive algorithms.

It is also possible to calculate Padé approximants on ascending staircases by means of a recursive computation scheme. To this end we formulate the following recurrence relations for which we again assume normality of the Padé table. First we introduce the notation

$$r_{m,n}(x) = \frac{\sum_{i=0}^{m} a_{m,n}^{(i)} \; x^i}{\sum_{i=0}^{n} b_{m,n}^{(i)} \; x^i}$$

Theorem 2.13.

If the Padé approximants

$$r_{m,n} = \frac{p_3}{q_3}, \quad r_{m,n-1} = \frac{p_2}{q_2} \quad \text{and} \quad r_{m+1,n-1} = \frac{p_1}{q_1},$$

are normal, then

$$\frac{p_3}{q_3} = \frac{a_{m,n-1}^{(m)} p_1 - a_{m+1,n-1}^{(m+1)} x p_2}{a_{m,n-1}^{(m)} q_1 - a_{m+1,n-1}^{(m+1)} x q_2}$$

Proof

It is easy to check that

$$\partial p_3 = \partial \left(a_{m,n-1}^{(m)} \; p_1 - a_{m+1,n-1}^{(m+1)} \; x p_2 \right) \leq m$$

$$\partial q_3 = \partial \left(a_{m,n-1}^{(m)} \; q_1 - a_{m+1,n-1}^{(m+1)} \; x q_2 \right) \leq n$$

$$\omega(f q_3 - p_3) \geq m + n + 1$$

Furthermore

$$\frac{p_3}{q_3} = r_{m,n}(x)$$

because of the assumption of normality and because of the uniqueness of the Padé approximant. ∎

This theorem enables us to calculate p_3/q_3 when p_1/q_1 and p_2/q_2 are given. We denote this by

$$\boxed{\begin{array}{l} r_{m,n-1} \\ \\ r_{m+1,n-1} \end{array}} \qquad \rightarrow \qquad r_{m,n}$$

Theorem 2.14.

If

$$r_{m-1,n} = \frac{p_3}{q_3}, \quad r_{m,n} = \frac{p_2}{q_2} \quad \text{and} \quad r_{m,n-1} = \frac{p_1}{q_1},$$

then

$$\frac{p_3}{q_3} = \frac{a_{m,n-1}^{(m)} p_2 - a_{m,n}^{(m)} p_1}{a_{m,n-1}^{(m)} q_2 - a_{m,n}^{(m)} q_1}$$

Computationally this means

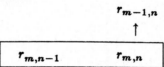

Combining theorem 2.13. and 2.14. we can compute the elements on an ascending staircase in the Padé table, starting with the first column.

Other algorithms exist for the computation of Padé approximants in a row, column or diagonal of the Padé table, instead of on staircases. We do not mention them here, but we refer to [1], [36] and [41].

3.7. The ϵ-algorithm.

Consider again a continued fraction

$$b_0 + \sum_{i=1}^{\infty} \frac{a_i}{\vert b_i}$$

with convergents

$$C_k = \frac{P_k}{Q_k}$$

Using theorem 1.2. of the previous chapter, we know that

$$\frac{1}{C_{k+1} - C_k} + \frac{1}{C_k - C_{k-1}}$$

$$= \frac{Q_{k+1} Q_k}{(-1)^{k+2} a_1 \ldots a_{k+1}} + \frac{Q_k Q_{k-1}}{(-1)^{k+1} a_1 \ldots a_k}$$

$$= \frac{Q_k}{(-1)^k a_1 \ldots a_k} \left[\frac{Q_{k+1} - a_{k+1} Q_{k-1}}{a_{k+1}} \right]$$

$$= \frac{b_{k+1} Q_k^2}{(-1)^k a_1 \ldots a_{k+1}}$$

$$= \frac{b_{k+1} b_{k+2} Q_k^2}{P_{k+2} Q_k - P_k Q_{k+2}}$$

For the continued fraction $g_{m-n}(x)$ given by (2.5b.) we get for $k = 2n$

$$\frac{1}{r_{m+1,n} - r_{m,n}} + \frac{1}{r_{m,n} - r_{m,n-1}} = \frac{Q_{2n}^2}{P_{2n+2} Q_{2n} - P_{2n} Q_{2n+2}}$$

and for $g_{m-n-1}(x)$ with $k = 2n + 1$ we have

$$\frac{1}{r_{m,n+1} - r_{m,n}} + \frac{1}{r_{m,n} - r_{m-1,n}} = \frac{Q_{2n}^2}{P_{2n+2} Q_{2n} - P_{2n} Q_{2n+2}}$$

Consequently the elements in a normal Padé table satisfy the relationship

$$(r_{m,n+1} - r_{m,n})^{-1} + (r_{m,n} - r_{m-1,n})^{-1} =$$

$$(r_{m+1,n} - r_{m,n})^{-1} + (r_{m,n} - r_{m,n-1})^{-1} \quad (2.8.)$$

where we have defined

$$r_{m,-1} = \infty$$
$$r_{-1,n} = 0$$

The identity (2.8.) is a star identity which relates

$$r_{m-1,n}(x) = N$$

$$r_{m,n-1}(x) = W \qquad r_{m,n}(x) = C \qquad r_{m,n+1}(x) = E$$

$$r_{m+1,n}(x) = S$$

and is often written as

$$(N - C)^{-1} + (S - C)^{-1} = (E - C)^{-1} + (W - C)^{-1}$$

If we introduce the following new notation for our Padé approximants

$$r_{m,n} = \epsilon_{2n}^{(m-n)}$$

we obtain a table of ϵ-values where again the subscript indicates a column and the superscript indicates a diagonal.

Table 2.6.

$\epsilon_0^{(0)}$	$\epsilon_2^{(-1)}$	$\epsilon_4^{(-2)}$	$\ldots$
$\epsilon_0^{(1)}$	$\epsilon_2^{(0)}$	$\epsilon_4^{(-1)}$	$\ldots$
$\epsilon_0^{(2)}$	$\epsilon_2^{(1)}$	$\epsilon_4^{(0)}$	$\ldots$
$\epsilon_0^{(3)}$	$\epsilon_2^{(2)}$	$\epsilon_4^{(1)}$	$\ldots$
$\vdots$	$\vdots$	$\vdots$	

The $\epsilon_0^{(m)}$ are the partial sums $F_m(x)$ of the Taylor series $f(x)$.

Remark the fact that only even column-indices occur. The table can be completed with odd-numbered columns in the following way. We define elements

$$\epsilon_{2n+1}^{(m-n-1)} = \epsilon_{2n-1}^{(m-n)} + \frac{1}{\epsilon_{2n}^{(m-n)} - \epsilon_{2n}^{(m-n-1)}} \qquad \begin{array}{l} m = 0, 1, \ldots \\ n = 0, 1, \ldots \end{array} \qquad (2.9.)$$

with

$$\begin{array}{ll} \epsilon_{-1}^{(m)} = 0 & m = 0, 1, \ldots \\ \epsilon_{2n}^{(-n-1)} = 0 & n = 0, 1, \ldots \end{array}$$

Table 2.7.

	$\epsilon_0^{(-1)}$		$\epsilon_2^{(-2)}$		$\epsilon_4^{(-3)}$	$\ldots$
$\epsilon_{-1}^{(0)}$		$\epsilon_1^{(-1)}$		$\epsilon_3^{(-2)}$		
	$\epsilon_0^{(0)}$		$\epsilon_2^{(-1)}$		$\epsilon_4^{(-2)}$	$\ldots$
$\epsilon_{-1}^{(1)}$		$\epsilon_1^{(0)}$		$\epsilon_3^{(-1)}$		
	$\epsilon_0^{(1)}$		$\epsilon_2^{(0)}$		$\epsilon_4^{(-1)}$	$\ldots$
$\epsilon_{-1}^{(2)}$		$\epsilon_1^{(1)}$		$\epsilon_3^{(0)}$		
	$\epsilon_0^{(2)}$		$\epsilon_2^{(1)}$		$\vdots$	
$\epsilon_{-1}^{(3)}$		$\epsilon_1^{(2)}$		$\vdots$		
	$\epsilon_0^{(3)}$		$\vdots$			
$\epsilon_{-1}^{(4)}$		$\vdots$				
	$\vdots$					
$\vdots$						

From the star identity we know that

$$\left(\epsilon_{2n+2}^{(m-n-1)} - \epsilon_{2n}^{(m-n)} \right)^{-1} + \left(\epsilon_{2n}^{(m-n)} - \epsilon_{2n}^{(m-n-1)} \right)^{-1} =$$

$$\left(\epsilon_{2n}^{(m-n+1)} - \epsilon_{2n}^{(m-n)} \right)^{-1} + \left(\epsilon_{2n}^{(m-n)} - \epsilon_{2n-2}^{(m-n+1)} \right)^{-1}$$

By (2.9.) this becomes

$$\left(\epsilon_{2n+2}^{(m-n-1)} - \epsilon_{2n}^{(m-n)} \right)^{-1} + \epsilon_{2n+1}^{(m-n-1)} - \epsilon_{2n-1}^{(m-n)} =$$

$$\epsilon_{2n+1}^{(m-n)} - \epsilon_{2n-1}^{(m-n+1)} + \left(\epsilon_{2n}^{(m-n)} - \epsilon_{2n-2}^{(m-n+1)} \right)^{-1}$$

from which we can easily conclude by induction on n that

$$\epsilon_{2n-1}^{(m-n+1)} = \epsilon_{2n-1}^{(m-n)} + \cfrac{1}{\epsilon_{2n}^{(m-n)} - \epsilon_{2n-2}^{(m-n+1)}}$$

or

$$\epsilon_{2n}^{(m-n)} = \epsilon_{2n-2}^{(m-n+1)} + \cfrac{1}{\epsilon_{2n-1}^{(m-n+1)} - \epsilon_{2n-1}^{(m-n)}} \qquad (2.10.)$$

The relations (2.9.) and (2.10.) are a means to calculate all the elements in table 2.7. and hence also to calculate all the Padé approximants in table 2.6. This algorithm is very handy when one needs the value of a Padé approximant for a given x and one does not want to compute the coefficients of the Padé approximant explicitly. The ε-algorithm was introduced in 1956 by Wynn [51]. To illustrate the procedure we calculate part of the completed ε-table for $f(x) = e^x$ with $x = 1$. Compare the obtained values with $e = 2.718281828\ldots$

Table 2.8.

$$f(x) = \exp(x)$$
$$x = 1$$

		0		0		0		0
0		1.000000		1.000000		1.500000		
	1.000000		∞		2.000000		3.000000	
0		1.000000		1.000000		2.500000		
	2.000000		3.000000		2.666667		2.727273	
0		2.000000		- 2.000000		19.00000		
	2.500000		2.750000		2.714286		2.718750	
0		6.000000		- 30.00000		243.0000		
	2.666667		2.722222		2.717949		2.718310	
0		24.00000		- 264.0000		3012.000		
	2.708333		2.718750		2.718254			
0		120.0000		-2280.000				
	2.716667		2.718333					
0		720.0000						
	2.718056							

Again computational difficulties can occur when the Padé table is not normal. Reformulations of the ε-algorithm in this case can be found in [15] and [52].

§4. Convergence of Padé approximants.

Let us consider a sequence $S = \{r_0, r_1, r_2, \ldots\}$ of elements from the Padé table for a given function $f(x)$. We want to investigate the existence of a function $F(x)$ with

$$\lim_{i \to \infty} r_i(x) = F(x)$$

and the properties of that function $F(x)$.

In general the convergence of S will depend on the properties of f. Before stating some general convergence results we give the following numerical examples. One can already remark that the poles of the elements in S will play an important role. A lot of information on the convergence of Padé approximants can also be found in [4].

4.1. Numerical examples.

For $f(x) = e^x$ and $r_i(x) = r_{m,n}(x)$ with $m + n = i$, we know [40 p. 246] that

$$\lim_{i \to \infty} r_i(x) = e^x \quad \text{for all } x \text{ in } \mathbb{C}$$

We illustrate this with the following numerical results.

Table 2.9.

$$f(x) = e^x$$
$$x = 1$$
$$e = 2.718281828\ldots$$

$m \backslash n$	0	1	2	3	4
0	1.000000	∞	2.000000	3.000000	2.666667
1	2.000000	3.000000	2.666667	2.727273	2.716981
2	2.500000	2.750000	2.714286	2.718750	2.718232
3	2.666667	2.722222	2.717949	2.718310	2.718280
4	2.708333	2.718750	2.718254	2.718284	2.718282

Next we consider the case that f is a rational function. For

$$f(x) = \frac{x + 10}{1 - x^2}$$

the Taylor series expansion

$$10 + x + 10x^2 + x^3 + 10x^4 + \dots$$

converges for $|x| < 1$. If $r_i(x) = r_{i,1}(x)$ then

$$r_i(x) = \sum_{k=0}^{i-1} c_k x^k + \frac{c_i x^i}{1 - \dfrac{c_{i+1}}{c_i} x}$$

For i even the pole of $r_i(x)$ is 10 and for i odd the pole of $r_i(x)$ is $\frac{1}{10}$. In these points the sequence $r_i(x)$ does not converge to $f(x)$.

For

$$f(x) = \frac{\ln(1+x)}{x} = 1 - \frac{x}{2} + \frac{x^2}{3} - \frac{x^3}{4} + \frac{x^4}{5} - \dots$$

the Taylor series expansion converges for $|x| < 1$ while the diagonal Padé approximants $r_{i,i}(x)$ converge to f for all x in $\mathbb{C} \backslash (-\infty, -1]$. The following results illustrate this.

Table 2.10.

$$f(x) = \frac{\ln(1+x)}{x}$$
$$f(1) = 0.69314718\dots$$
$$f(2) = 0.54930614\dots$$

i	$r_{i,i}(1)$	$r_{i,i}(2)$
0	1.000000	1.000000
1	0.700000	0.571429
2	0.693333	0.550725
3	0.693152	0.549403
4	0.693147	0.549313

4.2. Convergence of columns in the Padé table.

First we take $r_i(x) = r_{i,0}(x)$, the partial sums of the Taylor series expansion for $f(x)$. The following result is obvious.

Theorem 2.15.

If f is analytic in $B(0, r) = \{x \mid |x| < r\} \subseteq \mathbb{C}$ with $r > 0$, then $S = \{r_{i,0}\}_{i \in \mathbb{N}}$ converges uniformly to f on every closed and bounded subset of $B(0, r)$.

Next take $r_i(x) = r_{i,1}(x)$, the Padé approximants of order $(i, 1)$ for f. It is possible to construct functions f that are analytic in the whole complex plane but for which the poles of the $r_{i,1}$ are a dense subset of $\mathbb{C}$ [40 p. 158]. So in general S will not converge. But the following theorem can be proved [6].

Theorem 2.16.

If f is analytic in $B(0, r)$, then a subsequence of $\{r_{i,1}\}_{i \in \mathbb{N}}$ exists which converges uniformly to f on every closed and bounded subset of $B(0, r)$.

In [3] a similar result was proved for $S = \{r_{i,2}(x)\}_{i \in \mathbb{N}}$.
For meromorphic functions f it is also possible to prove the convergence of certain columns in the Padé table [22].

Theorem 2.17.

If f is analytic in $B(0, r)$ except in the poles $w_1, \ldots, w_k$ of f with total multiplicity n, then $\{r_{i,n}\}_{i \in \mathbb{N}}$ converges uniformly to f on every closed and bounded subset of $B(0, r) \backslash \{w_1, \ldots, w_k\}$.

4.3. Convergence of the diagonal elements.

In some cases a certain kind of convergence can be proved. It is called convergence in measure [39].

Theorem 2.18.

Let f be meromorphic and G a closed and bounded subset of $\mathbb{C}$. For every ϵ and δ in $\mathbb{R}_0^+$, there exists an integer k such that for $i > k$ we have

$$|r_{i,i}(x) - f(x)| < \epsilon \text{ for all } x \text{ in } G_i$$

where G_i is a subset of G such that the measure of $G \backslash G_i$ is less than δ.

Generalizations of this result can be found in [42].

4.4. Convergence of Padé approximants for Stieltjes series.

Stieltjes series were introduced in chapter I. For such series

$$f(z) = \sum_{i=0}^{\infty} d_i \, z^i$$

with convergence radius $\frac{1}{r}$ and

$$d_i = \int_0^r t^i dg(t)$$

where $g(t)$ is a real-valued, bounded, nondecreasing function taking on infinitely many different values, one can prove that the poles of the Padé approximants $r_{i+k,i}$ with $k \geq -1$ are simple and real positive. One obtains convergence of $\{r_{i,i}\}_{i\in\mathbb{N}}$ to an analytic function. The convergence is uniform on every closed and bounded subset of the cut complex plane $\mathbb{C}\backslash[\frac{1}{r}, \infty)$. Similar results hold for $\{r_{i+k,i}\}_{i\in\mathbb{N}}$ [8].

§5. Continuity of the Padé operator.

If $r_{m,n}$ is the Padé approximant of order (m, n) for f, then we call the operator $P_{m,n}$ that associates with f its (m, n) Padé approximant, the **Padé operator**. Here m and n are fixed. So we can adopt the notations

$$P = P_{m,n}$$
$$Pf = r_{m,n}$$

When we compute $r_{m,n}$ in finite precision arithmetic the computed result is not exactly the (m, n) Padé approximant, but it differs slightly from it by rounding errors and data perturbations. Since we can consider the computed result as the exact (m, n) Padé approximant of a slightly perturbed input power series, it is important to study the effect of such small perturbations on the operator P. To measure the small perturbations we introduce a pseudo-norm for formal power series:

$$\|c\| = \max_{0 \le i \le m+n} |c_i| \quad \text{with} \quad c = (c_0, \ldots, c_{m+n})$$

and the supremum norm for continuous functions on an interval $[a, b]$:

$$\|q\| = sup_{a \le x \le b} |q(x)|$$

The Padé approximants

$$r_{m,n}(x) = \frac{p_0}{q_0}(x)$$

were normalized such that $q_0(0) = 1$.
This implies the existence of an interval $[a, b]$ around the origin where $q_0(x)$ is strictly positive.
For given $f(x) = \sum_{i=0}^{\infty} c_i x^i$ we call a **neighbourhood** U_δ of f, the set of power series $g(x) = \sum_{i=0}^{\infty} d_i x^i$ such that $\|c - d\| \le \delta$. The following lemma is needed to prove the continuity of the operator P.

Lemma 2.2.

If Pf is normal, then a neighbourhood U_δ of f exists such that for all g in U_δ the approximant Pg is normal.

Proof

The lemma is an immediate consequence of theorem 2.6. by virtue of the fact that the determinants mentioned there are continuous functions of $c = (c_0, \ldots, c_{m+n})$. ∎

Theorem 2.19.

If $r_{m,n} = p_0/q_0$ is normal and q_0 is strictly positive in $[a, b]$, then there exist constants K and δ (only depending on $c = (c_0, \ldots, c_{m+n})$ and $[a, b]$) such that for every $g(x) = \sum_{i=0}^{\infty} d_i x^i$ with $\|c - d\| \leq \delta$:

$$\|Pf - Pg\| \leq K \|c - d\|$$

Proof

The fact that Pf is normal implies the existence of a neighbourhood U_ϵ of f such that for all g in U_ϵ the approximant Pg is normal. Hence, by theorem 2.11., a determinant formula for Pg can be given.
For

$$g(x) = \sum_{i=0}^{\infty} d_i x^i$$

we write

$$Pg = \frac{r_0}{s_0}(x) = \frac{\sum_{i=0}^{m} a_i(d_0, \ldots, d_{m+n})x^i}{\sum_{i=0}^{n} b_i(d_0, \ldots, d_{m+n})x^i} = \frac{\sum_{i=0}^{m} a_i(d)x^i}{\sum_{i=0}^{n} b_i(d)x^i}$$

where

$$Pf = \frac{p_0}{q_0}(x) = \frac{\sum_{i=0}^{m} a_i(c_0, \ldots, c_{m+n})x^i}{\sum_{i=0}^{n} b_i(c_0, \ldots, c_{m+n})x^i} = \frac{\sum_{i=0}^{m} a_i(c)x^i}{\sum_{i=0}^{n} b_i(c)x^i}$$

Since a_i and b_i are continously differentiable functions of c, a constant M exists such that

$$\|p_0 - r_0\| \leq \sup_{a \leq x \leq b} \sum_{i=0}^{m} |a_i(c) - a_i(d)| \, |x^i|$$

$$\leq M\|c - d\|$$

and

$$\|q_0 - s_0\| \leq sup_{a \leq x \leq b} \sum_{i=0}^{n} |b_i(c) - b_i(d)| \, |x^i|$$
$$\leq M\|c - d\|$$

for every power series g in U_ϵ. Write

$$\frac{p_0}{q_0} - \frac{r_0}{s_0} = \frac{(p_0 - r_0)q_0 + (s_0 - q_0)p_0}{q_0 s_0}$$

We know that $q_0(x)$ is strictly positive in $[a, b]$ and that $s_0(x)$ is a continuous function of d. Thus it is possible to construct a neighbourhood U_δ of f with $\delta \leq \epsilon$ and a constant N such that

$$\|\frac{1}{s_0}\| \leq N \quad \text{for all power series } g \text{ in } U_\delta$$

Now it is possible to bound

$$\|Pf - Pg\| \leq \|\frac{1}{q_0}\| \left(\|p\| + \|q\| \right) M \, N \, \|c - d\|$$
$$\leq K\|c - d\|$$

for all g in U_δ. ∎

So we have seen that in case of a normal approximant $r_{m,n}$, the Padé operator is continuous. Let's take a look at an example where the normality condition is not satisfied. Consider

$$f(x) = 1 + \alpha x + x^2$$

and take $m = 1 = n$. Then

$$r_{1,1}(x) = \frac{1 + \left(\alpha - \frac{1}{\alpha} \right) x}{1 - \frac{1}{\alpha} x} \quad \text{if } \alpha \neq 0$$

and

$$r_{1,1}(x) = 1 \quad \text{if} \quad \alpha = 0$$

If we write

$$r_\alpha(x) = \frac{\alpha + (\alpha^2 - 1)x}{\alpha - x}$$

and

$$r(x) = 1$$

then clearly for every x

$$\lim_{\alpha \to 0} r_\alpha(x) = r(x)$$

but

$$\lim_{\alpha \to 0} \|r_\alpha - r\| = \lim_{\alpha \to 0} \left(\sup_{a \le x \le b} |r_\alpha(x) - r(x)| \right) = \infty$$

for every interval $[a, b]$ around the origin.
However, a weakening of the normality condition is possible, in order to obtain a necessary and sufficient condition for the continuity of P.

Theorem 2.20.

The Padé operator P is continuous in f, if and only if $\min(m - m', n - n') = 0$ where m' and n' are the exact degrees of numerator and denominator of Pf respectively.

The proof can be found in [48].

§6. Multivariate Padé approximants.

6.1. Definition of multivariate Padé approximants.

We have seen in the previous sections that univariate Padé approximants can be obtained in several equivalent ways: one can solve the system of defining equations explicitly and thus obtain a determinant expression, one can set up a recursive scheme such as the ϵ-algorithm or one can construct a continued fraction whose convergents lie on a descending staircase in the Padé table. In the past few years all these approaches have been generalized to the multivariate case [12, 33, 34, 35, 37, 38, 45] but mostly the equivalence between the different techniques was lost. However, for the following definition a lot of properties of the univariate Padé approximant remain valid, also the recursive computation and the continued fraction representation.

We restrict ourselves to the case of two variables because the generalization for functions of more variables is straightforward. Consider the bivariate function $f(x, y)$ with Taylor series development

$$f(x, y) = \sum_{i,j=0}^{\infty} c_{ij}\ x^i\ y^j$$

around the origin. We know that a solution of the univariate Padé approximation problem (2.2.) for

$$f(x) = \sum_{i=0}^{\infty} c_i\ x^i$$

is given by

$$p(x) = \begin{vmatrix} \sum_{i=0}^{m} c_i x^i & x \sum_{i=0}^{m-1} c_i x^i & \ldots & x^n \sum_{i=0}^{m-n} c_i x^i \\ c_{m+1} & c_m & \ldots & c_{m+1-n} \\ \vdots & \vdots & \ddots & \vdots \\ c_{m+n} & c_{m+n-1} & \ldots & c_m \end{vmatrix}$$

and

$$q(x) = \begin{vmatrix} 1 & x & \cdots & x^n \\ c_{m+1} & c_m & \cdots & c_{m+1-n} \\ \vdots & \vdots & \ddots & \vdots \\ c_{m+n} & c_{m+n-1} & \cdots & c_m \end{vmatrix}$$

Let us now multiply the jth row in $p(x)$ and $q(x)$ by x^{m+j-1} $(j = 2, \ldots, n+1)$ and afterwards divide the jth column in $p(x)$ and $q(x)$ by x^{j-1} $(j = 2, \ldots, n+1)$. This results in a multiplication of numerator and denominator by x^{mn}. Having done so, we get

$$\frac{p(x)}{q(x)} = \frac{\begin{vmatrix} \sum_{i=0}^{m} c_i x^i & \sum_{i=0}^{m-1} c_i x^i & \cdots & \sum_{i=0}^{m-n} c_i x^i \\ c_{m+1} x^{m+1} & c_m x^m & \cdots & c_{m+1-n} x^{m+1-n} \\ \vdots & \vdots & \ddots & \vdots \\ c_{m+n} x^{m+n} & c_{m+n-1} x^{m+n-1} & \cdots & c_m x^m \end{vmatrix}}{\begin{vmatrix} 1 & 1 & \cdots & 1 \\ c_{m+1} x^{m+1} & c_m x^m & \cdots & c_{m+1-n} x^{m+1-n} \\ \vdots & \vdots & \ddots & \vdots \\ c_{m+n} x^{m+n} & c_{m+n-1} x^{m+n-1} & \cdots & c_m x^m \end{vmatrix}}$$

if $D = \det D_{m,n} \neq 0$.

This quotient of determinants can also immediately be written down for a bivariate function $f(x, y)$. The sum $\sum_{i=0}^{k} c_i x^i$ shall be replaced by the kth partial sum of the Taylor series development of $f(x, y)$ and the expression $c_k x^k$ by an expression that contains all the terms of degree k in $f(x, y)$. Here a bivariate term $c_{ij} x^i y^j$ is said to be of degree $i + j$.

If we define

$$
p(x,y) = \begin{vmatrix}
\sum_{i+j=0}^{m} c_{ij}x^i y^j & \sum_{i+j=0}^{m-1} c_{ij}x^i y^j & \cdots & \sum_{i+j=0}^{m-n} c_{ij}x^i y^j \\
\sum_{i+j=m+1} c_{ij}x^i y^j & \sum_{i+j=m} c_{ij}x^i y^j & \cdots & \sum_{i+j=m+1-n} c_{ij}x^i y^j \\
\vdots & & \ddots & \vdots \\
\sum_{i+j=m+n} c_{ij}x^i y^j & \sum_{i+j=m+n-1} c_{ij}x^i y^j & \cdots & \sum_{i+j=m} c_{ij}x^i y^j
\end{vmatrix}
$$

(2.11.)

and

$$
q(x,y) = \begin{vmatrix}
1 & 1 & \cdots & 1 \\
\sum_{i+j=m+1} c_{ij}x^i y^j & \sum_{i+j=m} c_{ij}x^i y^j & \cdots & \sum_{i+j=m+1-n} c_{ij}x^i y^j \\
\vdots & & \ddots & \vdots \\
\sum_{i+j=m+n} c_{ij}x^i y^j & \sum_{i+j=m+n-1} c_{ij}x^i y^j & \cdots & \sum_{i+j=m} c_{ij}x^i y^j
\end{vmatrix}
$$

(2.12.)

then it is easy to see that $p(x,y)$ and $q(x,y)$ are of the form

$$
p(x,y) = \sum_{i+j=mn}^{mn+m} a_{ij}x^i y^j
$$

$$
q(x,y) = \sum_{i+j=mn}^{mn+n} b_{ij}x^i y^j
$$

(2.13.)

Therefore we are interested in the following **multivariate Padé approximation problem**. Find bivariate polynomials p and q such that

$$
\begin{cases}
\omega p \geq mn & \partial p \leq mn + m \\
\omega q \geq mn & \partial q \leq mn + n \\
\omega[(fq - p)(x,y)] \geq mn + m + n + 1
\end{cases}
$$

(2.14.)

In comparison with the univariate problem (2.2.) the degrees of p and q and the order of p, q and $fq - p$ have been shifted over mn.

Theorem 2.21.

If $p(x, y)$ and $q(x, y)$ are given by (2.11.) and (2.12.) then

$$(fq - p)(x, y) = \sum_{i+j \geq mn+m+n+1} d_{ij}\, x^i\, y^j$$

Proof

If we put

$$A_k(x, y) = \sum_{i+j=mn+k} a_{ij}\, x_i\, y^j \qquad k = 0, \ldots, m$$

$$B_k(x, y) = \sum_{i+j=mn+k} b_{ij}\, x^i\, y^j \qquad k = 0, \ldots, n$$

$$C_k(x, y) = \sum_{i+j=k} c_{ij}\, x^i\, y^j \qquad k = 0, 1, 2, \ldots$$

then the conditions

$$(fq - p)(x, y) = \sum_{i+j \geq mn+m+n+1} d_{ij} x^i y^j$$

can be rewritten as follows :

$$\begin{cases} C_0(x, y)\, B_0(x, y) = A_0(x, y) \\ C_1(x, y)\, B_0(x, y) + C_0(x, y)\, B_1(x, y) = A_1(x, y) \\ \quad \vdots \\ C_m(x, y)\, B_0(x, y) + \ldots + C_{m-n}(x, y)\, B_n(x, y) = A_m(x, y) \end{cases} \qquad \text{(2.15a.)}$$

$$\begin{cases} C_{m+1}(x, y)\, B_0(x, y) + \ldots + C_{m+1-n}(x, y)\, B_n(x, y) = 0 \\ \quad \vdots \\ C_{m+n}(x, y)\, B_0(x, y) + \ldots + C_m(x, y)\, B_n(x, y) = 0 \end{cases} \qquad \text{(2.15b.)}$$

where $C_k = 0$ if $k < 0$. With this notation we have

$$p(x, y) = \sum_{k=0}^{m} A_k(x, y)$$

and

$$q(x, y) = \sum_{k=0}^{n} B_k(x, y)$$

If we solve the homogeneous system for the $B_k(x,y)$, choosing

$$B_0(x,y) = \begin{vmatrix} C_m(x,y) & \cdots & C_{m+1-n}(x,y) \\ \vdots & \ddots & \vdots \\ C_{m+n-1}(x,y) & \cdots & C_m(x,y) \end{vmatrix}$$

and using Cramer's rule, and if we substitute this solution in the system of equations defining the $A_k(x,y)$, then we get precisely $p(x,y)$ and $q(x,y)$ given by the determinant expressions (2.11.) and (2.12.). ∎

So $p(x,y)$ and $q(x,y)$ given by (2.11) and (2.12.) constitute a solution of (2.14.). Now the expressions (2.11.) and (2.12.) may be identically zero, but one can prove that problem (2.14.) always has at least one nontrivial solution [16]. For the definition of the (m,n) multivariate Padé approximant we first need the following property.

Theorem 2.22.

If p_1, q_1 and p_2, q_2 both satisfy the condition (2.14.) then

$$(p_1\, q_2)(x,y) = (p_2\, q_1)(x,y)$$

Proof

We proceed as in the univariate case. Write $p_1\, q_2 - p_2\, q_1$ as

$$(fq_2 - p_2)q_1 - (fq_1 - p_1)q_2$$

We know that

$$\omega(fq_1 - p_1) \geq mn + m + n + 1$$
$$\omega(fq_2 - p_2) \geq mn + m + n + 1$$
$$\omega q_1 \geq mn$$
$$\omega q_2 \geq mn$$

and consequently

$$\omega[(fq_2 - p_2)q_1 - (fq_1 - p_1)q_2] \geq 2mn + m + n + 1$$

Now

$$\partial(p_1 q_2 - p_2 q_1) \leq 2mn + m + n$$

So $p_1 q_2 - p_2 q_1$ is identically zero. ∎

The **multivariate Padé approximant** of order (m, n) for $f(x, y)$ is now defined as the irreducible form

$$r_{m,n}(x, y) = \frac{p_0}{q_0}(x, y)$$

of a rational function $p(x, y)/q(x, y)$ where p and q satisfy (2.14.). For these $r_{m,n}$ a lot of properties of univariate Padé approximants remain valid.

6.2. Block structure of the multivariate Padé table.

From the previous section we easily conclude the following.

Theorem 2.23.

For every m and n a unique multivariate Padé approximant of order (m, n) for $f(x, y)$ exists.

Let us now first take a look at an example. Consider

$$f(x, y) = 1 + \frac{x}{0.1 - y} + \sin(xy)$$

$$= 1 + \sum_{i=1}^{\infty} 10^i\, xy^{i-1} + \sum_{i=0}^{\infty} (-1)^i \frac{(xy)^{2i+1}}{(2i + 1)!}$$

$$= 1 + 10x + 101xy + 1000xy^2 + \ldots$$

Take $m = 1 = n$. Then we have to look for $p(x, y)$ and $q(x, y)$ of the form

$$p(x, y) = a_{10}x + a_{01}y + a_{20}x^2 + a_{11}xy + a_{02}y^2$$

$$q(x, y) = b_{10}x + b_{01}y + b_{20}x^2 + b_{11}xy + b_{02}y^2$$

such that

$$(f\,q - p)(x, y) = \sum_{i+j \geq 4} d_{ij}\, x^i\, y^j$$

According to theorem 2.21. a solution is given by

$$p(x, y) = \begin{vmatrix} 1 + 10x & 1 \\ 101xy & 10x \end{vmatrix}$$

$$= 10x + 100x^2 - 101xy$$

$$q(x, y) = \begin{vmatrix} 1 & 1 \\ 101xy & 10x \end{vmatrix}$$

$$= 10x - 101xy$$

$$r_{1,1}(x, y) = \frac{1 + 10x - 10.1y}{1 - 10.1y}$$

Here the shift of the degrees over $mn = 1$ has disappeared by taking the irreducible form of

$$\frac{p}{q}(x, y) = \frac{10x + 100x^2 - 101xy}{10x - 101xy}$$

This is not always the case. Take $m = 1$ and $n = 2$. Then $p(x, y)$ and $q(x, y)$ satisfying (2.14.) are

$$p(x, y) = \begin{vmatrix} 1 + 10x & 1 & 0 \\ 101xy & 10x & 1 \\ 1000xy^2 & 101xy & 10x \end{vmatrix}$$

$$= 100x^2 - 101xy + 1000x^3 - 2020x^2y + 1000xy^2$$

$$q(x, y) = \begin{vmatrix} 1 & 1 & 1 \\ 101xy & 10x & 1 \\ 1000xy^2 & 101xy & 10x \end{vmatrix}$$

$$= 100x^2 - 101xy - 1010x^2y + 1000xy^2 + 201x^2y^2$$

We obtain

$$r_{1,2}(x,y) = \frac{x - 1.01y + 10x^2 - 20.2xy + 10y^2}{x - 1.01y - 10.1xy + 10y^2 + 2.01xy^2}$$

In general the following results can be proved about the order and degree of numerator and denominator of $r_{m,n}(x,y)$.

Since p_0/q_0 is the irreducible form of a rational function p/q with p and q of the form (2.13.), we may write

$$p(x,y) = p_0(x,y)\, T(x,y)$$

$$q(x,y) = q_0(x,y)\, T(x,y)$$

with

$$\omega q_0 \le \omega q$$

and because of (2.15a.) also with

$$\omega q_0 \le \omega p_0$$

So we can define

$$m' = \partial p_0 - \omega q_0$$

$$n' = \partial q_0 - \omega q_0$$

Obviously

$$
\begin{aligned}
m' &= \partial p_0 - \omega q_0 = (\partial p - \partial T) - \omega q_0 \\
&\le \partial p - \omega T - \omega q_0 = \partial p - \omega q \le mn + m - mn = m \\
n' &= \partial q_0 - \omega q_0 = (\partial q - \partial T) - \omega q_0 \\
&\le \partial q - \omega T - \omega q_0 = \partial q - \omega q \le mn + n - mn = n
\end{aligned}
$$

This definition of m' and n' is an extension of the univariate definition, because in the univariate case $\omega q_0 = 0$.

Theorem 2.24.

If the Padé approximant of order (m, n) for $f(x, y)$ is given by

$$r_{m,n}(x, y) = \frac{p_0}{q_0}(x, y)$$

then an integer s with $0 \leq s \leq \min(m - m', n - n')$ and a homogeneous bivariate polynomial

$$S(x, y) = \sum_{i+j=mn-\omega q_0+s} s_{ij} \, x^i \, y^j$$

exist such that $p(x, y) = S(x, y) \, p_0(x, y)$ and $q(x, y) = S(x, y) \, q_0(x, y)$ satisfy (2.14.).

Proof

Since

$$\frac{p_0}{q_0}(x, y)$$

is computed from a solution of (2.14.), we may consider nontrivial polynomials $p_1(x, y)$ and $q_1(x, y)$ and write

$$p_1(x, y) = T(x, y) \, p_0(x, y)$$

$$q_1(x, y) = T(x, y) \, q_0(x, y)$$

with p_1 and q_1 satisfying (2.14.) and with

$$T(x, y) = \sum_{i+j=\omega T}^{\partial T} t_{ij} \, x^i \, y^j$$

a bivariate polynomial.
Clearly

$$\omega q_1 = \omega T + \omega q_0 \geq mn$$

and thus

$$\omega T = mn - \omega q_0 + s$$

with $s \geq 0$.

Also

$$\omega T \leq \partial T$$

with

$$\partial T = \partial p_1 - \partial p_0 \leq mn + m - (m' + \omega q_0) = mn - \omega q_0 + m - m'$$
$$\partial T = \partial p_1 - \partial p_0 \leq mn + n - (n' + \omega q_0) = mn - \omega q_0 + n - n'$$

Hence

$$\partial T \leq mn - \omega q_0 + \min(m - m', n - n')$$

which implies

$$0 \leq s \leq \min(m - m', n - n')$$

Write

$$S(x, y) = \sum_{i+j=mn-\omega q_0+s} t_{ij}\, x^i\, y^j$$

Because

$$mn + m + n + 1 \leq \omega(f q_1 - p_1) = \omega[(f q_0 - p_0)T]$$
$$= \omega[(f q_0 - p_0)S]$$

the proof is completed. ∎

Theorem 2.25.

If the Padé approximant of order (m, n) for $f(x, y)$ is given by

$$r_{m,n}(x, y) = \frac{p_0}{q_0}(x, y)$$

and $m' = \partial p_0 - \omega q_0$ and $n' = \partial q_0 - \omega q_0$, then:

(a) $\omega(f q_0 - p_0) = \omega q_0 + m' + n' + t + 1$ with $t \geq 0$

(b) for k and ℓ satisfying $m' \leq k \leq m' + t$ and $n' \leq \ell \leq n' + t$:
$r_{k,\ell}(x, y) = r_{m,n}(x, y)$

(c) $m \leq m' + t$ and $n \leq n' + t$

The proof is based on the univariate version and can be found in [19, 20]. From this theorem it must be clear that normality can be defined exactly in the same way as in the univariate case. Necessary and sufficient conditions for normality of $r_{m,n}(x, y)$ are given in problem (9) at the end of this chapter. As a conclusion we take a closer look at the meaning of the numbers ωq_0, m' and n'.

In the solution $p(x, y)$ and $q(x, y)$ the degrees have been shifted over mn. By taking the irreducible form of p/q part of that shift can disappear, but what remains in p_0, q_0 and $fq_0 - p_0$ is a shift over ωq_0. Now m' and n' play the same role as in the univariate case: they measure the exact degree of a polynomial by disregarding the shift over ωq_0.

6.3. The multivariate ε-algorithm.

Because of the theorems 2.21. and 2.22. and using the notation

$$C_k(x, y) = \sum_{i+j=k} c_{ij} x^i y^j$$

we can write for a normal bivariate Padé approximant

$$r_{m,n}(x, y) = \frac{\begin{vmatrix} \sum_{k=0}^{m} C_k(x,y) & \sum_{k=0}^{m-1} C_k(x,y) & \cdots & \sum_{k=0}^{m-n} C_k(x,y) \\ C_{m+1}(x,y) & C_m(x,y) & \cdots & C_{m+1-n}(x,y) \\ \vdots & \vdots & \ddots & \vdots \\ C_{m+n}(x,y) & C_{m+n-1}(x,y) & \cdots & C_m(x,y) \end{vmatrix}}{\begin{vmatrix} 1 & 1 & \cdots & 1 \\ C_{m+1}(x,y) & C_m(x,y) & \cdots & C_{m+1-n}(x,y) \\ \vdots & \vdots & \ddots & \vdots \\ C_{m+n}(x,y) & C_{m+n-1}(x,y) & \cdots & C_m(x,y) \end{vmatrix}}$$

Using this determinant formula one can prove [17] that with

$$\epsilon_0^{(m)} = \sum_{i+j=0}^{m} c_{ij} \, x^i \, y^j = \sum_{k=0}^{m} C_k(x, y)$$

which are the partial sums of the bivariate Taylor series $f(x, y)$, we have

$$\epsilon_{2n}^{(m-n)} = r_{m,n}(x, y)$$

if the ϵ-table is calculated using the formulas (2.9.) and (2.10.).

Remark the analogy with the univariate theory: in both cases the $\epsilon_0^{(m)}$ are the mth partial sums and the same algorithm is used.

The reformulation of the ϵ-algorithm in case the multivariate Padé table is not normal is also inspired by the univariate results [20].

6.4. The multivariate qd-algorithm.

First we rewrite the univariate qd-algorithm in a form such that it can immediately be generalized. Consider the function

$$f(x) = \sum_{i=0}^{\infty} c_i\, x^i$$

and its Padé approximant of order (m, n). If $r_{m,n}(x)$ is the $(2n)^{th}$ convergent of the continued fraction

$$\sum_{i=0}^{m-n} c_i\, x^i + \left.\frac{c_{m-n+1}\, x^{m-n+1}}{1}\right| - \left.\frac{q_1^{(m-n+1)}\, x}{1}\right| - \left.\frac{e_1^{(m-n+1)}\, x}{1}\right| -\ldots$$

with, for $k \geq 1$ and $\ell \geq 1$

$$q_1^{(k)} = \frac{c_{k+1}}{c_k}$$

$$e_0^{(k)} = 0$$

$$e_\ell^{(k)} = e_{\ell-1}^{(k+1)} + q_\ell^{(k+1)} - q_\ell^{(k)}$$

$$q_{\ell+1}^{(k)} = q_\ell^{(k+1)} \frac{e_\ell^{(k+1)}}{e_\ell^{(k)}}$$

then we can also say that $r_{m,n}(x)$ is the $(2n)^{th}$ convergent of the continued fraction

$$\sum_{i=0}^{m-n} c_i\, x^i + \left.\frac{c_{m-n+1}\, x^{m-n+1}}{1}\right| - \left.\frac{Q_1^{(m-n+1)}}{1}\right| - \left.\frac{E_1^{(m-n+1)}}{1}\right| -\ldots$$

with

$$Q_1^{(k)} = \frac{c_{k+1}\, x^{k+1}}{c_k\, x^k}$$

$$E_0^{(k)} = 0$$

$$E_\ell^{(k)} = E_{\ell-1}^{(k+1)} + Q_\ell^{(k+1)} - Q_\ell^{(k)}$$

$$Q_{\ell+1}^{(k)} = Q_\ell^{(k+1)} \frac{E_\ell^{(k+1)}}{E_\ell^{(k)}}$$

We have simply included the factor x in $Q_\ell^{(k)}$ and $E_\ell^{(k)}$.
This last continued fraction can be generalized for a bivariate function in the same way as the formulas (2.11.) and (2.12.) were obtained: replace the expression $c_k\, x^k$ by an expression that contains all the terms of degree k in the bivariate series

$$\sum_{i+j=0}^{\infty} c_{ij}\, x^i\, y^j$$

We define, for $k \geq 1$ and $\ell \geq 1$

$$Q_1^{(k)}(x,y) = \frac{\displaystyle\sum_{i+j=k+1} c_{ij}\, x^i\, y^j}{\displaystyle\sum_{i+j=k} c_{ij}\, x^i\, y^j}$$

$$E_0^{(k)}(x,y) = 0$$

$$E_\ell^{(k)}(x,y) = E_{\ell-1}^{(k+1)}(x,y) + Q_\ell^{(k+1)}(x,y) - Q_\ell^{(k)}(x,y)$$

$$Q_{\ell+1}^{(k)}(x,y) = Q_\ell^{(k+1)}(x,y) \frac{E_\ell^{(k+1)}(x,y)}{E_\ell^{(k)}(x,y)}$$

In [18] one proves that now $r_{m,n}(x,y)$ is the $(2n)^{th}$ convergent of the continued fraction

$$\sum_{i+j=0}^{m-n} c_{ij} x^i y^j + \left. \frac{\sum_{\substack{i+j=\\m-n+1}} c_{ij}\, x^i\, y^j}{1} \right| - \left. \frac{Q_1^{(m-n+1)}(x,y)}{1} \right| - \left. \frac{E_1^{(m-n+1)}(x,y)}{1} \right| - \cdots$$

$$(2.16.)$$

We give a simple example. Consider

$$f(x,y) = \frac{xe^x - ye^y}{x - y} = \sum_{i,j=0}^{\infty} \frac{x^i y^j}{(i+j)!}$$

$$= 1 + x + y + \frac{1}{2}(x^2 + xy + y^2) + \dots$$

Take $m = 2$ and $n = 1$. The Padé approximant $r_{2,1}(x,y)$ is given by

$$r_{2,1}(x,y) = \frac{p_0}{q_0}(x,y)$$

with

$$p_0(x,y) = \begin{vmatrix} 1 + x + y + \frac{1}{2}(x^2 + xy + y^2) & 1 + x + y \\ \frac{1}{6}(x^3 + x^2y + xy^2 + y^3) & \frac{1}{2}(x^2 + xy + y^2) \end{vmatrix}$$

$$= \frac{1}{2}(x^2 + xy + y^2) + \frac{1}{3}(x^3 + \frac{5}{2}x^2y + \frac{5}{2}xy^2 + y^3)$$
$$+ \frac{1}{12}(x^4 + 2x^3y + 5x^2y^2 + 2xy^3 + y^4)$$

$$q_0(x,y) = \begin{vmatrix} 1 & 1 \\ \frac{1}{6}(x^3 + x^2y + xy^2 + y^3) & \frac{1}{2}(x^2 + xy + y^2) \end{vmatrix}$$

$$= \frac{1}{2}(x^2 + xy + y^2) - \frac{1}{6}(x^3 + x^2y + xy^2 + y^3)$$

Indeed $r_{2,1}(x,y)$ is the second convergent of the continued fraction (2.16.):

$$r_{2,1}(x,y) = 1 + x + y + \frac{\frac{1}{2}(x^2 + xy + y^2)|}{|\ 1\ } - \frac{\frac{x^3 + x^2y + xy^2 + y^3}{3(x^2 + xy + y^2)}|}{|\ 1\ }$$

where the qd-scheme was started with

$$Q_1^{(k)}(x,y) = \frac{x^{k+1} + x^k y + \dots + xy^k + y^{k+1}}{(k+1)(x^k + x^{k-1}y + \dots + xy^{k-1} + y^k)}$$

Problems.

(1)   a) Show that the normalization $q_0(0) = 1$ is always possible for

$$r_{m,n}(x) = \frac{p_0}{q_0}(x)$$

b) Show that $p_0(0) = c_0$ if $q_0(0) = 1$.

(2)   Let $f(x)$ be analytic. The rational function

$$r_{m,n}(x) = \frac{p_0}{q_0}(x)$$

is the Padé approximant of order (m, n) for f if and only if

$$f^{(k)}(0) = r_{m,n}^{(k)}(0) \quad \text{for} \quad k = 0, \ldots, m + n + j$$

with j the largest possible integer $(j \in \mathbb{Z})$.

(3)   Give an explicit formula for $r_{1,1}(x)$ using the Taylor coefficients of $f(x)$.

(4)   Let

$$r_{m,n}(x) = \frac{p_0}{q_0}(x)$$

be the Padé approximant of order (m, n) for f. If $f(0) \neq 0$ then

$$\frac{\gamma q_0}{\gamma p_0}(x)$$

is the Padé approximant of order (m, n) for $\frac{1}{f}$ where $\gamma = \frac{1}{f}(0)$.
This property is called the **reciprocal covariance** of Padé approximants.

(5)   a) Let $r_{m,m}(x)$ be the Padé approximant of order (m, m) for f.
        If $c\, f(0) + d \neq 0$ then
$$\frac{a + b\, r_{m,m}(x)}{c + d\, r_{m,m}(x)}$$

is the Padé approximant of order (m, m) for

$$\frac{a + bf}{c + df}$$

This property is called the **homografic covariance** of Padé approximants.

b) Is this property also valid for off-diagonal approximants?

(6) Prove theorem 2.6.

(7) Prove theorem 2.14.

(8) Let $r_{m,m}(x)$ be the Padé approximant of order (m, m) for f. Then

$$r_{m,m}\left(\frac{ax}{1 + bx}\right)$$

is the Padé approximant of order (m, m) for

$$f\left(\frac{ax}{1 + bx}\right)$$

(9) Prove the following conditions for normality of a multivariate Padé approximant:

$$r_{m,n}(x, y) = \frac{p_0}{q_0}(x, y)$$

is normal if and only if

$$m' = \partial p_0 - \omega q_0 = m$$
$$n' = \partial q_0 - \omega q_0 = n$$
$$\omega(f q_0 - p_0) = \omega q_0 + m + n + 1$$

(10) Formulate and prove the reciprocal and homografic covariance of multivariate Padé approximants.

Remarks.

(1)   Condition (2.2.) is a linear condition and the polynomials $p(x)$ and $q(x)$
        satisfying (2.2.) are not necessarily irreducible. Sometimes the Padé ap-
        proximant of order (m, n) for f is defined in a non-linear way: let p/q be
        an irreducible rational function with numerator and denominator respec-
        tively of degree at most m and n, then p/q is called the Padé approximant
        of order (m, n) for f if [2]

$$\omega(f - \frac{p}{q}) \geq m + n + 1$$

        If this definition is used it is obvious that the Padé approximation problem
        does not always have a solution.

(2)   We have here used the functions $\{x^i\}_{i \in \mathbb{N}}$ as a basis for the polynomials.
        Instead one could also use a set of orthogonal polynomials. In this way
        one can e.g. study the Legendre-Padé approximation problem [26, 27] or
        the Chebyshev-Padé approximation problem [14, 25, 28].
        More information about orthogonal polynomials and their link with the
        Padé approximation problem can be found in [10]: almost all the recursive
        methods for computing sequences of Padé approximants can be derived
        from the theory of formal orthogonal polynomials. A general subroutine
        for the computation of Padé approximants is given in [10] for the normal
        case and has been developed by Draux and van Ingelandt in the non-
        normal case.

(3)   Padé approximants have also been defined for matrix-valued functions
        [5, 46], for the operator exponential [24], for formal power series in
        a parameter with non-commuting elements of a certain algebra as
        coefficients [7, 23], for vector valued functions [21] and so on.

(4)   Although Henri Padé was not the creator of the so-called Padé ap-
        proximants, his name was given to those approximants because of the
        extensive study of their properties in his thesis in 1892. Brezinski has
        edited a book containing the works of Padé and the story of his life [9].

(5)   Padé approximants are in fact a special case of so-called **Padé-type
        approximants** [10]. There a rational function p/q with numerator and
        denominator respectively of degree at most m and n is computed such
        that

$$\omega(f q - p) \geq m + 1$$

        This supplies us with a linear system of $m + 1$ equations in $m + n + 1$

unknowns. The remaining $n + 1$ free parameters are used for a normalization and to insert some extra information about f if it is available, for instance some knowledge about singularities of f.

References.

[1] *Baker G.*   Recursive calculation of Padé approximants. In [31], 83-92.

[2] *Baker G.*   Essentials of Padé approximants. Academic Press, New York, 1975.

[3] *Baker G.* and *Graves-Morris P.*   Convergence of the Padé table. J. Math. Anal. Appl. 57, 1977, 323-339.

[4] *Baker G.* and *Graves-Morris P.*   Padé Approximants: Basic Theory. Encyclopedia of Mathematics and its applications: vol 13. Addison-Wesley, Reading, 1981.

[5] *Basu S.* and *Bose N.*   Two-dimensional matrix Padé approximants: existence, nonuniqueness and recursive computation. IEEE Trans. Automat. Control 25, 1980, 509-514.

[6] *Beardon A.*   The convergence of Padé approximants. J. Math. Anal. Appl. 21, 1968, 344-346.

[7] *Bessis J.* and *Talman J.*   Variational approach to the theory of operator Padé approximants. Rocky Mountain J. Math. 4, 1974, 151-158.

[8] *Brezinski C.*   Approximants de Padé et applications. Cours de questions spéciales de Math. I (UCL), Belgium, 1981.

[9] *Brezinski C.*   Henri Padé: Œuvres. Librairie Scient. et Techn. Blanchard, Paris, 1984.

[10] *Brezinski C.*.   Padé type approximation and general othogonal polynomials. ISNM 50, Birkhäuser Verlag, Basel, 1980.

[11] *Bultheel A.*   Recursive algorithms for the Padé table: two approaches. In [49], 211-230.

[12] *Chisholm J.*   N-variable rational approximants. In [44], 23-42.

[13] *Claessens G.* and *Wuytack L.*   On the computation of non-normal Padé approximants. J. Comput. Appl. Math. 5, 1979, 283-289.

[14] *Clenshaw C.* and *Lord K.*   Rational approximations from Chebyshev series. In [44], 95-113.

[15] *Cordellier F.*   Démonstration algébrique de l'extension de l'identité de Wynn aux tables de Padé non normales. In [49], 36-60.

[16] *Cuyt A.* Multivariate Padé approximants. J. Math. Anal. Appl. 96, 1983, 283-293.

[17] *Cuyt A.* The ϵ-algorithm and multivariate Padé approximants. Numer. Math. 40, 1982, 39-46.

[18] *Cuyt A.* The qd-algorithm and multivariate Padé approximants. Numer. Math. 42, 1983, 259-269.

[19] *Cuyt A.* Padé approximants for operators: theory and applications. Lecture Notes in Mathematics 1065, Springer, Berlin, 1984.

[20] *Cuyt A.* Singular rules for the calculation of non-normal multivariate Padé approximants. J. Comput. Appl. Math. **14, 1986, 289-301**

[21] *de Bruin M.* Generalized C-fractions and a multidimensional Padé table. Ph. D., University of Amsterdam, 1974.

[22] *de Montessus de Ballore R.* Sur les fractions continues algébriques. Rend. Circ. Mat. Palermo 19, 1905, 1-73.

[23] *Draux A.* The Padé approximants in a non-commutative algebra and their applications. In [47], 117-131.

[24] *Fair W.* and *Luke Y.* Padé approximants to the operator exponential. Numer. Math. 14, 1970, 379-382.

[25] *Fike C.* Computer evaluation of Mathematical functions. Prentice Hall, New Jersey, 1968, 181-190.

[26] *Fleischer J.* Analytic continuation of scattering amplitudes and Padé approximants. Nuclear Phys. B 37, 1972, 59-76.

[27] *Fleischer J.* Nonlinear Padé approximants for Legendre series. J. Math. Phys. 14, 1973, 246-248.

[28] *Gragg W.* Laurent-, Fourier- and Chebyshev-Padé tables. In [44], 61-72.

[29] *Gragg W.* The Padé table and its relation to certain algorithms of numerical analysis. SIAM Rev. 14, 1972, 1-62.

[30] *Gragg W.* Matrix interpretations and applications of the continued fraction algorithm. Rocky Mountain J. Math. 4, 1974, 213-225.

[31] *Graves-Morris P.*   Padé approximants and their applications. Academic Press, London, 1973.

[32] *Henrici P.*   Applied and computational complex analysis: vol. 1 &2. John Wiley, New York, 1976.

[33] *Hughes Jones R.*   General rational approximants in n variables. J. Approx. Theory 16, 1976, 201-233.

[34] *Karlsson J.* and *Wallin H.*   Rational approximation by an interpolation procedure in several variables. In [44], 83-100.

[35] *Levin D.*   General order Padé type rational approximants defined from double power series. J. Inst. Math. Appl. 18, 1976, 1-8.

[36] *Longman I.*   Computation of the Padé table. Internat. J. Comput. Math. 3, 1971, 53-64.

[37] *Lutterodt C.*   Rational approximants to holomorphic functions in n dimensions. J. Math. Anal. Appl. 53, 1976, 89-98.

[38] *Murphy J. and O'Donohoe M.*   A two-variable generalization of the Stieltjes-type continued fraction. J. Comput. Appl. Math. 4, 1978, 181-190.

[39] *Nuttall J.*   The convergence of Padé approximants of meromorphic functions. J. Math. Anal. Appl. 31, 1970, 147-153.

[40] *Perron O.*   Die Lehre von den Kettenbruchen II. Teubner, Stuttgart, 1977.

[41] *Pindor M.*   A simplified algorithm for calculating the Padé table derived from Baker and Longman schemes. J. Comp. Appl. Math. 2, 1976, 255-258.

[42] *Pommerenke Ch.*   Padé approximants and convergence in capacity. J. Math. Anal. Appl. 41, 1973, 775-780.

[43] *Rutishauser H.*   Der Quotienten-Differenzen Algorithmus. Mitteilungen Institut für angewandte Mathematik (ETH) 7, Birkhäuser Verlag, Basel, 1957.

[44] *Saff E. and Varga R.*   Padé and rational approximation : theory and applications. Academic Press, London, 1977.

[45] *Sciafe B.*   Studies in Numerical Analysis. Academic Press, London, 1972.

[46] *Starkand Y.*   Explicit formulas for matrix-valued Padé approximants. J. Comput. Appl. Math. 5, 1979, 63-66.

[47] *Werner H.* and *Bünger H.*   Padé approximation and its applications. Lecture Notes in Mathematics 1071, Springer, Berlin, 1984.

[48] *Werner H.* and *Wuytack L.*   On the continuity of the Padé operator. SIAM J. Numer. Anal. 20, 1983, 1273-1280.

[49] *Wuytack L.*   Padé approximation and its applications. Lecture Notes in Mathematics 765, Springer, Berlin, 1979.

[50] *Wuytack L.*   Commented bibliography on techniques for computing Padé approximants. In [49], 375-392.

[51] *Wynn P.*   On a device for computing the $e_m(S_n)$ transformation. MTAC 10, 1956, 91-96.

[52] *Wynn P.*   Singular rules for certain nonlinear algorithms. BIT 3, 1963, 175-195.

CHAPTER III: Rational Interpolants.

"Die Lagrangesche Interpolationsformel, welche dazu dient, eine Reihe von n Werthen durch eine ganze Funktion (n-1)^{ten} Grades darzustellen, ist von Cauchy durch eine Formel verallgemeinert worden, welche eine Reihe von n+m Werthen durch eine gebrochne Funktion darstellt, deren Zähler und Nenner respektive von (n-1)^{ten} und m^{ten} Grade sind. Man kann die Formel dadurch deduciren, dass man die lineären Gleichungen, von welchen die Aufgabe abhängt, auflöst, und die Determinanten, welche man für den Zähler und Nenner findet, entwickelt."

C. JACOBI — *"Über die Darstellung einer Reihe gegebner Werthe durch eine gebrochne rationale Funktion" (1845).*

§1. Notations and definitions.

Consider a function f defined on a subset G of the complex plane. Let $\{x_i\}_{i\in\mathbb{N}}$ be a sequence of different points belonging to G. We still denote the exact degree of a polynomial p by ∂p and its order by ωp. The **rational interpolation problem** of order (m, n) for f consists in finding polynomials

$$p(x) = \sum_{i=0}^{m} a_i x^i$$

and

$$q(x) = \sum_{i=0}^{n} b_i x^i$$

with $p(x)/q(x)$ irreducible and such that

$$f(x_i) = \frac{p}{q}(x_i) \qquad\qquad i = 0, \ldots, m+n \qquad\qquad (3.1.)$$

Instead of solving problem (3.1.) we consider the linear system of equations

$$f(x_i)q(x_i) - p(x_i) = 0 \qquad\qquad i = 0, \ldots, m+n \qquad\qquad (3.2.)$$

Condition (3.2.) is a homogeneous system of $m + n + 1$ linear equations in the $m + n + 2$ unknown coefficients a_i and b_i of p and q. Hence the system (3.2.) always has at least one nontrivial solution. For different solutions of (3.2.) the following equivalence can be proved.

Theorem 3.1.

If the polynomials p_1, q_1 and p_2, q_2 both satisfy (3.2.) then $p_1 q_2 = p_2 q_1$.

Proof

For the polynomial $p_1 q_2 - p_2 q_1$ we can write

$$(p_1 q_2 - p_2 q_1)(x_i) = [(fq_2 - p_2)q_1 - (fq_1 - p_1)q_2](x_i) = 0 \qquad i = 0, \ldots, m+n$$

Since $\partial(p_1 q_2 - p_2 q_1) \leq m + n$ it must vanish identically for it has more than $m + n$ zeros.   ∎

Not all solutions of (3.2.) also satisfy (3.1.): it is very well possible that the polynomials p and q satisfying (3.2.) are such that p/q is reducible. Nevertheless, because of theorem 3.1., all solutions of (3.2.) have the same irreducible form. For p and q satisfying (3.2.) we shall denote by

$$r_{m,n}(x) = \frac{p_0}{q_0}(x)$$

the irreducible form of p/q, where $q_0(x)$ is normalized such that $q_0(x_0) = 1$, and we shall call $r_{m,n}(x)$ the **rational interpolant** of order (m, n) for f.
The following result is a consequence of theorem 3.1.

Theorem 3.2.

For every nonnegative m and n a unique rational interpolant of order (m, n) for f exists.

Although the terminology "interpolant" is used it may be that $r_{m,n}(x)$ does not satisfy the interpolation conditions (3.1.) anymore [14]. A simple example will illustrate this. Let $x_0 = 0$, $x_1 = 1$, $x_2 = 2$ and $f(x_0) = 0$, $f(x_1) = 3$, $f(x_2) = 3$. Take $m = n = 1$. Then the system of interpolation conditions is

$$\begin{cases} a_0 = 0 \\ 3(b_0 + b_1) - (a_0 + a_1) = 0 \\ 3(b_0 + 2b_1) - (a_0 + 2a_1) = 0 \end{cases}$$

A solution is $p(x) = 3x$ and $q(x) = x$. Thus $p_0(x) = 3$ and $q_0(x) = 1$. Clearly

$$\frac{p_0}{q_0}(x_0) \neq f(x_0)$$

Note the similarity with the Padé approximation problem: the Padé approximant of order (m, n) did not necessarily satisfy condition (2.2.) anymore. We shall see that many properties valid for Padé approximants can be generalized for rational interpolants [20].

§2. Fundamental properties.

2.1. Properties of the rational interpolant.

Let $r_{m,n} = p_0/q_0$ be the rational interpolant of order (m,n) for f. If p_0 and q_0 do not satisfy the system of conditions (3.2) themselves, it is easy to construct polynomials p and q from p_0 and q_0 that are a solution of (3.2.). Denote the exact degree of p_0 by $\mathbf{m'}$ and the exact degree of q_0 by $\mathbf{n'}$.

Theorem 3.3.

If the rational interpolant of order (m,n) for f is

$$r_{m,n}(x) = \frac{p_0}{q_0}(x)$$

then an integer s exists with $0 \leq s \leq \min(m - m', n - n')$ and s points $y_1, \ldots, y_s$ exist belonging to $\{x_0, \ldots, x_{m+n}\}$ such that

$$p(x) = p_0(x) \prod_{i=1}^{s}(x - y_i)$$

and

$$q(x) = q_0(x) \prod_{i=1}^{s}(x - y_i)$$

satisfy (3.2.).

Proof

Let $p_1(x)$ and $q_1(x)$ be a solution of (3.2.). Then

$$(f\, q_1)(x_i) = p_1(x_i) \quad i = 0, \ldots, m + n$$

Hence if $q_1(x_i) = 0$ also $p_1(x_i) = 0$. Let $\{y_1, \ldots, y_s\}$ be the set of zeros of $q_1(x)$ belonging to $\{x_0, \ldots, x_{m+n}\}$. We construct

$$t(x) = \prod_{i=1}^{s}(x - y_i)$$

If

$$r_{m,n}(x) = \frac{p_0}{q_0}(x)$$

then a polynomial v exists with

$$p_1(x) = v(x)\, t(x)\, p_0(x)$$

$$q_1(x) = v(x)\, t(x)\, q_0(x)$$

Consequently

$$s = \partial t \leq min(m - m', n - n')$$

Since

$$q_1(x_i) \neq 0 \text{ for } x_i \in \{x_0, \ldots, x_{m+n}\} \backslash \{y_1, \ldots, y_s\}$$

we have

$$(fq_0)(x_i) = p_0(x_i) \quad \text{for} \quad x_i \in \{x_0, \ldots, x_{m+n}\} \backslash \{y_1, \ldots, y_s\}$$

If we put $p(x) = p_0(x)\, t(x)$ and $q(x) = q_0(x)\, t(x)$ then p and q also satisfy (3.2.). ∎

As a conclusion we can say that the rational interpolation problem (3.1.) has a solution if and only if $p_0(x)$ and $q_0(x)$ satisfy the system of equations (3.2.).

2.2. The table of rational interpolants.

The rational interpolants of order (m, n) for f can again be ordered in a table:

$r_{0,0}$	$r_{0,1}$	$r_{0,2}$	$\ldots$
$r_{1,0}$	$r_{1,1}$	$r_{1,2}$	$\ldots$
$r_{2,0}$	$r_{2,1}$	$\ldots$	
$r_{3,0}$	$r_{3,1}$	$\ldots$	
$\vdots$	$\vdots$		

In the first column one finds the polynomial interpolants for f and in the first row the inverses of the polynomial interpolants for $\frac{1}{f}$. By theorem 3.3. at least

$m' + n' + t + 1$ points $\{z_0, \ldots, z_{m'+n'+t}\}$ with $t \geq 0$ exist in $\{x_0, \ldots, x_{m+n}\}$ such that

$$r_{m,n}(z_i) = f(z_i) \quad i = 0, \ldots, m' + n' + t$$

On the basis of this conclusion a property comparable with the block structure of the Padé table can be formulated.

Theorem 3.4.

Let

$$r_{m,n}(x) = \frac{p_0}{q_0}(x)$$

and let

(a) $(f\, q_0 - p_0)(x_i) = 0 \quad i = 0, \ldots, m' + n' + t$

(b) $(f\, q_0 - p_0)(x_i) \neq 0 \quad i = m' + n' + t + 1, \ldots, m + n$

then (c)   for k and ℓ satisfying $m' \leq k \leq m' + t$ and $n' \leq \ell \leq n' + t$:

$$r_{k,\ell}(x) = \frac{p_0}{q_0}(x)$$

(d) $m \leq m' + t$ and $n \leq n' + t$

Proof

For k and ℓ satisfying $m' \leq k \leq m' + t$ and $n' \leq \ell \leq n' + t$ we define $s = min(k - m', \ell - n')$ and

$$t(x) = \prod_{i=m'+n'+t+1}^{m'+n'+t+s} (x - x_i)$$

For $p(x) = p_0(x)\, t(x)$ and $q(x) = q_0(x)\, t(x)$ we know from (a) that

$$(f\, q - p)(x_i) = 0 \quad i = 0, \ldots, m' + n' + t + s$$

Since $k + \ell \leq m' + n' + t + s$ we have

$$(f\, q - p)(x_i) = 0 \quad i = 0, \ldots, k + \ell$$

and thus

$$r_{k,\ell} = \frac{p_0}{q_0}$$

because

$$\partial\, p \leq k$$
$$\partial q \leq l$$

Since p_0/q_0 is the irreducible form of p/q, where p and q satisfy

$$(f\, q - p)(x_i) = 0 \quad i = 0,\ldots, m+n$$

it is possible, according to theorem 3.3., to find an integer s with $0 \leq s \leq min(m - m', n - n')$ such that

$$m + n \leq m' + n' + t + s$$

The upper bound on s implies the upper bounds $m' + t$ and $n' + t$ on m and n respectively. This finishes the proof. ∎

Remark that condition (b) in theorem 3.4. is only necessary to assure that the integer t in (a) is as large as possible. It is important to emphasize that the table of rational interpolants has the block structure described above, only when the interpolation points $\{x_0, \ldots, x_{m+n}\}$ are ordered in such a fashion that the points x_i in which

$$(f\, q - p)(x_i) \neq 0$$

are the last points in the interpolation set. For chosen m and n such a numbering can always be arranged and then we know that all the rational interpolants lying in a square emanating from $r_{m',n'}$ and with $r_{m'+t,n'+t}$ as its furthermost corner are equal to p_0/q_0.

Let us take a look at the following example.

Let $x_i = i$ for $i = 0,\ldots,5$ with $f(x_0) = 1$, $f(x_1) = 1$, $f(x_2) = 5/3$, $f(x_3) = 5/2$, $f(x_4) = 17/5$, $f(x_5) = 13/5$. Then

$$r_{3,2}(x) = \frac{p_0(x)}{q_0(x)} = \frac{1 + x^2}{1 + x}$$

and $(f \, q_0 - p_0)(x_i) = 0$ for $i = 0, \dots, 4$, while $(f \, q_0 - p_0)(x_5) \neq 0$. This implies that $r_{2,1} = r_{2,2} = r_{3,1} = r_{3,2}$.

If the interpolation points are not numbered such that the conditions of theorem 3.4. are satisfied then it is possible for 2 elements in the table of rational interpolants to be equal without having a square of equal elements. A disturbance in the numbering of the interpolation points in theorem 3.4. implies a disturbance in the block-structure. Let us illustrate this.

For $x_i = i$ when $i = 0, \dots, 3$ and $f(x_0) = 2$, $f(x_1) = 3/2$, $f(x_2) = 4/5$, $f(x_3) = 1/2$ we find

$$r_{1,0}(x) = r_{2,1}(x) = 2 - \frac{x}{2}$$

but

$$r_{2,0}(x) = 2 - \frac{2}{5}x - \frac{1}{10}x^2$$

and

$$r_{1,1}(x) = \frac{2 - 5x/7}{1 - x/7}$$

One can also check that the rational interpolation problem of order (k, ℓ) for f as formulated in (3.1.) has a solution if the integers k and ℓ are such that

$$m' \leq k$$

$$n' \leq \ell$$

$$m' + n' + t \geq k + \ell$$

and if the conditions of theorem 3.4. are satisfied.

2.3. Normality.

Again we call an entry of the table **normal** if it occurs only once in that table. A necessary condition for the normality of the rational interpolant $r_{m,n}(x)$ is formulated in the following theorem.

Theorem 3.5.

If

$$r_{m,n} = \frac{p_0}{q_0}$$

is normal and if $(f\, q_0 - p_0)(x_i) = 0$ for $i = 0, \ldots, m' + n'$, then

    (a) $m = m'$ and $n = n'$

    (b) $(fq_0 - p_0)(x_i) \neq 0$ for $i = m + n + 1, m + n + 2$

Proof

Since

$$(f\, q_0 - p_0)(x_i) = 0 \quad \text{for} \quad i = 0, \ldots, m' + n'$$

we know that $r_{m,n} = r_{m',n'}$. Hence, if $r_{m,n}$ is normal, we have $m = m'$ and $n = n'$.

Now suppose that $(f\, q_0 - p_0)(x_{m+n+1}) = 0$. Then

$$r_{m,n} = r_{m+1,n} = r_{m,n+1}$$

This contradicts the normality of $r_{m,n}$.

If $(f\, q_0 - p_0)(x_{m+n+2}) = 0$ then

$$p(x) = (x - x_{m+n+1})p_0(x)$$

and

$$q(x) = (x - x_{m+n+1})q_0(x)$$

satisfy

$$(f\, q - p)(x_i) = 0 \quad \text{for} \quad i = 0, \ldots, m + n + 2$$

Thus $r_{m,n} = r_{m+1,n+1}$ which is again a contradiction with the normality of $r_{m,n}$.   ■

Conclusion (b) in theorem 3.5. does not imply that

$$(fq_0 - p_0)(x_i) \neq 0 \quad \text{for} \quad i \geq m + n + 1$$

That the conditions (a) and (b) are not sufficient to guarantee normality of $r_{m,n}(x)$ is illustrated as follows. Let $x_i = i$ for $i = 0, 1, 2, \ldots$ and $f(x_0) = 0$, $f(x_1) = 1$, $f(x_2) = 3$, $f(x_3) = 4$, $f(x_i) = i$ for $i = 4, 5, 6, \ldots$. For $m = 1$ and $n = 0$ we find $r_{m,n}(x) = x$ with (a) and (b) in theorem 3.5. satisfied. But $r_{m,n}$ is not normal because $r_{m,n} = r_{k,\ell}$ for $k \geq 3$ and $\ell \geq 2$. However, it is possible to formulate a sufficient condition for the normality of $r_{m,n}$.

Theorem 3.6.

If

$$r_{m,n} = \frac{p_0}{q_0}$$

with $m = m'$, $n = n'$ and $(fq_0 - p_0)(x_i) = 0$ for at most $m + n + 1$ points from the sequence $\{x_i\}_{i \in \mathbb{N}}$, then $r_{m,n}$ is normal.

Proof

Let us suppose that $r_{m,n}$ is not normal, in other words that $r_{m,n} = r_{k,\ell}$ for $k \geq m$, $\ell \geq n$ and with $k + \ell > m + n$. According to theorem 3.3. an integer s exists with $0 \leq s \leq \min(k - m, \ell - n)$ and s points $\{y_1, \ldots, y_s\}$ exist such that the polynomials

$$p(x) = p_0(x) \prod_{i=1}^{s} (x - y_i)$$

and

$$q(x) = q_0(x) \prod_{i=1}^{s} (x - y_i)$$

satisfy

$$(fq - p)(x_i) = 0 \quad \text{for} \quad i = 0, \ldots, k + \ell$$

Hence $(fq_0 - p_0)(x_i) = 0$ for at least $k + \ell + 1 - s$ points in $\{x_i\}_{i \in \mathbb{N}}$. Since s is bounded above by $k - m$ and $\ell - n$, we conclude that $k + \ell + 1 - s > m + n + 1$ which contradicts the fact that $(fq_0 - p_0)(x_i) = 0$ for at most $m + n + 1$ points from $\{x_i\}_{i \in \mathbb{N}}$ ∎

§3. Methods to compute rational interpolants.

In the sequel of this chapter we suppose that every rational interpolant $r_{m,n}(x)$ itself satisfies the interpolation conditions (3.1.). This is for instance satisfied if $\min(m - m', n - n') = 0$.

3.1. Interpolating continued fractions.

Theorem 3.7.

If

$$r_{m,n} = \frac{p_1}{q_1}$$

and

$$r_{m+k,n+\ell} = \frac{p_2}{q_2}$$

with $k, \ell \geq 0$ then a polynomial $v(x)$ exists with

$$\partial v \leq \max(k - 1, \ell - 1)$$

and

$$(p_1 q_2 - p_2 q_1)(x) = v(x) B_{m+n+1}(x)$$

where

$$B_{m+n+1}(x) = \prod_{i=0}^{m+n} (x - x_i)$$

Proof

As we assumed, the rational functions $r_{m,n}$ and $r_{m+k,n+\ell}$ both satisfy (3.1.):

$$(fq_1 - p_1)(x_i) = 0 \quad \text{for} \quad i = 0, \ldots, m + n$$
$$(fq_2 - p_2)(x_i) = 0 \quad \text{for} \quad i = 0, \ldots, m + k + n + \ell$$

Consequently

$$(p_1 q_2 - p_2 q_1)(x_i) = [(fq_2 - p_2)q_1](x_i) - [(fq_1 - p_1)q_2](x_i) = 0$$

for $i = 0, \ldots, m + n$ and thus a polynomial $v(x)$ exists such that $(p_1 q_2 - p_2 q_1)(x) = v(x) B_{m+n+1}(x)$. It is easy to see that $\partial v \leq \max(k - 1, \ell - 1)$ since $\partial(p_1 q_2 - p_2 q_1) \leq \max(m + n + k, m + n + \ell)$. ∎

If we consider the staircase of rational interpolants

$$T_k = \{r_{k,0}, r_{k+1,0}, r_{k+1,1}, r_{k+2,1}, \ldots\} \quad \text{for} \quad k \geq 0$$

it is possible to compute coefficients $d_i (i \geq 0)$ such that the convergents of the continued fraction

$$d_0 + d_1(x - x_0) + \ldots + d_k(x - x_0)\ldots(x - x_{k-1})$$

$$+ \left|\frac{d_{k+1}(x - x_0)\ldots(x - x_k)}{1}\right| + \left|\frac{d_{k+2}(x - x_{k+1})}{1}\right| + \left|\frac{d_{k+3}(x - x_{k+2})}{1}\right| + \ldots$$

$$(3.3.)$$

are precisely the subsequent elements of T_k.

Theorem 3.8.

If every three consecutive elements in T_k are different, then a continued fraction of the form (3.3.) exists with $d_{k+i} \neq 0$ for $i \geq 1$ and such that the nth convergent equals the $(n + 1)^{th}$ element of T_k.

The proof is left to the reader because it is completely analogous to the one given for theorem 2.9. We shall now describe methods that can be used to calculate those coefficients $d_i (i \geq 0)$.

3.2. Inverse differences.

Inverse differences for a function f given in G are defined as follows:

$$\varphi_0[x] = f(x) \quad \text{for every} \quad x \text{ in } G$$

$$\varphi_1[x_0, x_1] = \frac{x_1 - x_0}{\varphi_0[x_1] - \varphi_0[x_0]} \quad \text{for every} \quad x_0, x_1 \text{ in } G$$

$$\varphi_k[x_0, x_1, \ldots, x_{k-2}, x_{k-1}, x_k] =$$

$$\frac{x_k - x_{k-1}}{\varphi_{k-1}[x_0, \ldots, x_{k-2}, x_k] - \varphi_{k-1}[x_0, \ldots, x_{k-2}, x_{k-1}]}$$

$$\text{for every} \quad x_0, x_1, \ldots, x_k \text{ in } G$$

We call $\varphi_k[x_0,\ldots,x_k]$ the kth **inverse difference** of f in the points $x_0,\ldots,x_k$. Usually inverse differences depend on the numbering of the points $x_0,\ldots,x_k$ although they are independent of the order of the last two points. If we want to calculate an interpolating continued fraction of the form

$$d_0 + \left.\frac{x-x_0}{d_1}\right| + \left.\frac{x-x_1}{d_2}\right| + \left.\frac{x-x_2}{d_3}\right| +\ldots \qquad (3.4.)$$

we have to compute the inverse differences in table 3.1.

Table 3.1.

$\varphi_0[x_0]$				
$\varphi_0[x_1]$	$\varphi_1[x_0,x_1]$			
$\varphi_0[x_2]$	$\varphi_1[x_0,x_2]$	$\varphi_2[x_0,x_1,x_2]$		
$\varphi_0[x_3]$	$\varphi_1[x_0,x_3]$	$\varphi_2[x_0,x_1,x_3]$		
$\vdots$	$\vdots$	$\vdots$		
$\varphi_0[x_n]$	$\varphi_1[x_0,x_n]$	$\varphi_2[x_0,x_1,x_n]$	$\ldots$	$\varphi_n[x_0,\ldots,x_n]$

Theorem 3.9.

If $d_i = \varphi_i[x_0,\ldots,x_i]$ in the continued fraction (3.4.), then the nth convergent C_n of (3.4.) satisfies

$$C_n(x_i) = f(x_i) \text{ for } i = 0,\ldots,n$$

if $C_n(x_i)$ is defined.

Proof

From the definition of inverse differences we know that for $n \geq 1$:

$$f(x) = \varphi_0[x]$$
$$= \varphi_0[x_0] + \frac{x - x_0}{\varphi_1[x_0, x]}$$
$$= \varphi_0[x_0] + \cfrac{x - x_0}{\varphi_1[x_0, x_1] + \cfrac{x - x_1}{\varphi_2[x_0, x_1, x]}}$$
$$= \varphi_0[x_0] + \left.\frac{x - x_0}{\varphi_1[x_0, x_1]}\right| + \left.\frac{x - x_1}{\varphi_2[x_0, x_1, x_2]}\right|$$
$$+ \ldots + \left.\frac{x - x_{n-1}}{\varphi_n[x_0, \ldots, x_{n-1}, x]}\right|$$

With $d_i = \varphi_i[x_0, \ldots, x_i]$ it follows that C_n satisfies the imposed interpolation conditions. ∎

The continued fraction

$$\varphi_0[x_0] + \left.\frac{x - x_0}{\varphi_1[x_0, x_1]}\right| + \left.\frac{x - x_1}{\varphi_2[x_0, x_1, x_2]}\right| + \ldots$$

is called a **Thiele interpolating continued fraction**. To illustrate this technique we give the following example. Consider the data: $x_i = i$ for $i = 0, \ldots, 3$, $f(x_0) = 1$, $f(x_1) = 3$, $f(x_2) = 2$ and $f(x_3) = 4$. We get

1			
3	1/2		
2	2	2/3	
4	1	4	3/10

The rational function

$$1 + \left.\frac{x}{1/2}\right| + \left.\frac{x - 1}{2/3}\right| + \left.\frac{x - 2}{3/10}\right| = \frac{5x^2 - 5x - 6}{4x - 6} = r(x)$$

indeed satisfies $r(x_i) = f(x_i)$ for $i = 0, \ldots, 3$.

In the previous example difficulties occured neither for the computation of the inverse differences nor for the evaluation of $r(x_i)$. We shall illustrate the existence of such computational difficulties by means of some examples. Consider again the data: $x_0 = 0$, $x_1 = 1$, $x_2 = 2$ with $f(x_0) = 0$, $f(x_1) = 3 = f(x_2)$. Then the table of inverse differences looks like

$$0$$

$$3 \qquad 1/3$$

$$3 \qquad 2/3 \qquad 3$$

Hence

$$r(x) = 0 + \left.\frac{x}{}\right|_{1/3} + \left.\frac{x-1}{}\right|_{3} = \frac{3x}{x}$$

is not defined for $x = x_0$, and thus we cannot guarantee the satisfaction of the interpolation condition $r(x_0) = f(x_0)$.

If we consider the data: $x_i = i$ for $i = 0, \ldots, 4$ with $f(x_0) = 1$, $f(x_1) = 0$, $f(x_2) = 2$, $f(x_3) = -2$ and $f(x_4) = 5$ then $\varphi_2[x_0, x_1, x_3]$ is not finite. This does not imply the nonexistence of the rational interpolant in question. A simple permutation of the interpolation data enables us to continue the computations. For $x_0 = 0$, $x_1 = 2$, $x_2 = 1$, $x_3 = 3$ and $x_4 = 4$ we get

$$1$$

$$2 \qquad 2$$

$$0 \qquad -1 \qquad 1/3$$

$$-2 \qquad -1 \qquad -1/3 \qquad -3$$

$$5 \qquad 1 \qquad -2 \qquad -9/7 \qquad 7/12$$

The rational function

$$r(x) = 1 + \left.\frac{x}{}\right|_{2} + \left.\frac{x-2}{}\right|_{1/3} + \left.\frac{x-1}{}\right|_{-3} + \left.\frac{x-3}{}\right|_{7/12} = \frac{23x^2 - 85x + 62}{12x^2 - 59x + 62}$$

satisfies $r(x_i) = f(x_i)$ for $i = 0, \ldots, 4$.

In order to avoid this dependence upon the numbering of the data we will introduce reciprocal differences.

3.3. Reciprocal differences.

Reciprocal differences for a function f given in G are defined as follows :

$$\rho_0[x] = f(x) \text{ for every } x \text{ in } G$$

$$\rho_1[x_0, x_1] = \frac{x_1 - x_0}{\rho_0[x_1] - \rho_0[x_0]} \text{ for every } x_0, x_1 \text{ in } G$$

$$\rho_2[x_0, x_1, x_2] = \frac{x_2 - x_1}{\rho_1[x_0, x_2] - \rho_1[x_0, x_1]} + \rho_0[x_0] \text{ for every } x_0, x_1, x_2 \text{ in } G$$

$$\rho_k[x_0, \ldots, x_k] = \frac{x_k - x_{k-1}}{\rho_{k-1}[x_0, \ldots, x_{k-2}, x_k] - \rho_{k-1}[x_0, \ldots, x_{k-2}, x_{k-1}]}$$

$$+ \rho_{k-2}[x_0, \ldots, x_{k-2}]$$

$$\text{for every } x_0, \ldots, x_k \text{ in } G$$

We call $\rho_k[x_0, \ldots, x_k]$ the kth **reciprocal difference** of the function f in the points $x_0, \ldots, x_k$. There is a close relationship between inverse and reciprocal differences as stated in the next property.

Theorem 3.10.

For $k \geq 2$ and for all $x_0, \ldots, x_k$ in G:

$$\varphi_0[x_0] = \rho_0[x_0]$$
$$\varphi_1[x_0, x_1] = \rho_1[x_0, x_1]$$
$$\varphi_k[x_0, \ldots, x_k] = \rho_k[x_0, \ldots, x_k] - \rho_{k-2}[x_0, \ldots, x_{k-2}]$$

Proof

The relations above are an immediate consequence of the definitions. ∎

This theorem is helpful for the proof of the following important property.

Theorem 3.11.

$\rho_k[x_0, \ldots, x_k]$ does not depend upon the numbering of the points $x_0, \ldots, x_k$.

Proof

We consider the continued fraction (3.4.) and calculate the kth convergent by means of the recurrence relations (1.3.):

$$P_{-1} = 1 \qquad P_0 = d_0$$
$$Q_{-1} = 0 \qquad Q_0 = 1$$

$$\begin{cases} P_i = d_i\, P_{i-1} + (x - x_{i-1})P_{i-2} \\[2mm] Q_i = d_i\, Q_{i-1} + (x - x_{i-1})Q_{i-2} \end{cases} \qquad i = 1, \ldots, k$$

For even $k = 2j$, this convergent is of the form

$$\frac{a_0 + a_1 x + \ldots + a_j\, x^j}{b_0 + b_1 x + \ldots + b_j\, x^j}$$

and for odd $k = 2j - 1$, it is of the form

$$\frac{a_0 + a_1 x + \ldots + a_j\, x^j}{b_0 + b_1 x + \ldots + b_{j-1}\, x^{j-1}}$$

In both cases we calculate the coefficients of the terms of highest degree in numerator and denominator, using the recurrence relations for the kth convergent and the previous theorem :
for k even we get

$$a_j = \rho_k[x_0, \ldots, x_k]$$
$$b_j = 1$$

and for k odd

$$a_j = 1$$
$$b_{j-1} = \rho_k[x_0, \ldots, x_k]$$

Since $\rho_k[x_0, \ldots, x_k]$ appears to be a quotient of coefficients in the rational interpolant, it is independent of the ordering of the $x_0, \ldots, x_k$ because the rational interpolant itself is independent of that ordering. ∎

The interpolating continued fraction of the form (3.4.) can now also be calculated as follows: compute a table of reciprocal differences and put $d_0 = \rho_0[x_0]$, $d_1 = \rho_1[x_0, x_1]$ and for $i \geq 2$: $d_i = \rho_i[x_0, \ldots, x_i] - \rho_{i-2}[x_0, \ldots, x_{i-2}]$.

Up to now we have only constructed rational interpolants lying on the descending staircase T_0. To calculate a rational interpolant on T_k with $k > 0$ one proceeds as follows. Obviously it is possible to construct a continued fraction of the form

$$c_0 + c_1(x - x_0) + \ldots + c_k(x - x_0)\ldots(x - x_{k-1})$$

$$+ \left. \frac{c_{k+1}(x - x_0)\ldots(x - x_k)}{1} \right| + \left. \frac{x - x_{k+1}}{d_{k+2}} \right| + \left. \frac{x - x_{k+2}}{d_{k+3}} \right| + \ldots$$

$$(3.5.)$$

whose convergents are the elements of T_k.

Clearly $c_0, \ldots, c_{k+1}$ are the divided differences $f[x_0], \ldots, f[x_0, \ldots, x_{k+1}]$ since $r_{k,0}$ and $r_{k+1,0}$, the first two convergents, are the polynomial interpolants for f of degree k and $k + 1$ respectively.

If we want to calculate for instance $r_{k+\ell,\ell}$ we need the $(2\ell)^{th}$ convergent of (3.5.). In order to compute the coefficients d_{k+i} for $i = 2, \ldots, 2\ell$ we write

$$r_{k+\ell,\ell}(x) = p(x) + \frac{q(x)}{s(x)}$$

where

$$p(x) = c_0 + c_1(x - x_0) + \ldots + c_k(x - x_0)\ldots(x - x_{k-1})$$
$$q(x) = c_{k+1}(x - x_0)\ldots(x - x_k)$$

To define s we proceed as follows. The conditions

$$r_{k+\ell,\ell}(x_i) = f(x_i) \quad \text{for} \quad i = 0, \ldots, k + 2\ell$$

imply that s must satisfy

$$s(x_i) = \frac{q(x_i)}{f(x_i) - p(x_i)} \quad \text{for} \quad i = k + 1, \ldots, k + 2\ell$$

So $s(x)$ is the $(2\ell - 1)^{th}$ convergent of the continued fraction

$$1 + \left. \frac{x - x_{k+1}}{d_{k+2}} \right| + \left. \frac{x - x_{k+2}}{d_{k+3}} \right| + \ldots$$

Hence $s(x)$ belongs to the descending staircase T_0 in the table of rational interpolants for the function

$$\frac{q}{f - p}$$

As soon as the coefficients $c_0, \ldots, c_{k+1}$ are known, the function $q/(f - p)$ can be constructed and inverse or reciprocal differences for it can be computed. The coefficients d_{k+i} with $i \geq 2$ are precisely those inverse differences. So finally the computation of an element in T_k for f is reduced to the computation of an element in T_0 for

$$\frac{q}{f - p}$$

3.4. A generalization of the qd-algorithm.

Consider continued fractions of the form

$$g_k(x) = c_0 + \sum_{i=1}^{k} c_i(x - x_0)(x - x_1)\ldots(x - x_{i-1})$$

$$+ \left| \frac{c_{k+1}(x - x_0)\ldots(x - x_k)}{1} \right| + \left| \frac{-q_1^{(k+1)}(x - x_{k+1})}{1 + q_1^{(k+1)}(x_0 - x_{k+1})} \right|$$

$$+ \left| \frac{-e_1^{(k+1)}(x - x_{k+2})}{1 + e_1^{(k+1)}(x_0 - x_{k+2})} \right| + \left| \frac{-q_2^{(k+1)}(x - x_{k+3})}{1 + q_2^{(k+1)}(x_0 - x_{k+3})} \right|$$

$$+ \left| \frac{-e_2^{(k+1)}(x - x_{k+4})}{1 + e_2^{(k+1)}(x_0 - x_{k+4})} \right| + \ldots \qquad (3.6.)$$

Theorem 3.12.

If every three consecutive elements in T_k are different, then a continued fraction of the form (3.6.) exists with $c_{k+1} \neq 0$, $q_i^{(k+1)} \neq 0$, $e_i^{(k+1)} \neq 0$, $1 + q_i^{(k+1)}(x_0 - x_{k+2i-1}) \neq 0$, $1 + e_i^{(k+1)}(x_0 - x_{k+2i}) \neq 0$ for $i \geq 1$ and such that the nth convergent equals the $(n + 1)^{th}$ element of T_k.

Proof

For the elements in T_k we put

$$r_{k+i,j} = \frac{p_{k+i,j}}{q_{k+i,j}} \text{ for } i = j, j+1 \text{ and } j = 0, 1, 2, \ldots$$

and for the convergents of $g_k(x)$ we put

$$C_n = \frac{P_n}{Q_n} \text{ for } n = 0, 1, 2, \ldots \text{ with } Q_0 = Q_1 = 1$$

Using theorem 1.4. a continued fraction with n^{th} convergent equal to

$$\frac{P_n}{Q_n} \quad (n \geq 0)$$

is, after an equivalence transformation,

$$P_0 + \frac{P_1 - P_0|}{|1} + \sum_{i=1}^{\infty} \frac{\left|\dfrac{P_i Q_{i+1} - P_{i+1} Q_i}{P_i Q_{i-1} - P_{i-1} Q_i}\right.}{\left.\dfrac{P_{i+1} Q_{i-1} - P_{i-1} Q_{i+1}}{P_i Q_{i-1} - P_{i-1} Q_i}\right|} \qquad (3.7.)$$

For

$$C_n = \frac{P_n}{Q_n} = r_{k+\left\lfloor \frac{n+1}{2} \right\rfloor, \left\lfloor \frac{n}{2} \right\rfloor} = \frac{p_{k+\left\lfloor \frac{n+1}{2} \right\rfloor, \left\lfloor \frac{n}{2} \right\rfloor}}{q_{k+\left\lfloor \frac{n+1}{2} \right\rfloor, \left\lfloor \frac{n}{2} \right\rfloor}} \quad n = 0, 1, 2, \ldots$$

we find by means of theorem 3.7. that for $i \geq 1$,

$$\frac{P_i Q_{i+1} - P_{i+1} Q_i}{P_i Q_{i-1} - P_{i-1} Q_i} = a_i(x - x_{k+i})$$

$$\frac{P_{i+1} Q_{i-1} - P_{i-1} Q_{i+1}}{P_i Q_{i-1} - P_{i-1} Q_i} = b_i$$

with $a_i \neq 0$ and $b_i \neq 0$. The continued fraction (3.7.) is then

$$c_0 + c_1(x - x_0) + \ldots + c_k(x - x_0) \ldots (x - x_{k-1})$$

$$+ \frac{c_{k+1}(x - x_0) \ldots (x - x_k)|}{|1} + \sum_{i=1}^{\infty} \frac{a_i(x - x_{k+i})|}{|b_i}$$

with $c_{k+1} \neq 0$ since $r_{k,0} \neq r_{k+1,0}$.

For (3.7.) it is even true that

$$P_n = p_{k+\lfloor \frac{n+1}{2} \rfloor, \lfloor \frac{n}{2} \rfloor}$$
$$Q_n = q_{k+\lfloor \frac{n+1}{2} \rfloor, \lfloor \frac{n}{2} \rfloor}$$

Hence

$$1 = Q_n(x_0) = b_n Q_{n-1}(x_0) + a_n(x_0 - x_{k+n})Q_{n-2}(x_0) = b_n + a_n(x_0 - x_{k+n})$$

or

$$b_n = 1 - a_n(x_0 - x_{k+n}) \quad n \geq 0$$

If we put for $i \geq 1$

$$a_{2i-1} = -q_i^{(k+1)}$$
$$a_{2i} = -e_i^{(k+1)}$$

we find that (3.7.) can be written as (3.6.).   ∎

To calculate the coefficients $q_i^{(k+1)}$ and $e_i^{(k+1)}$ in (3.6.) one can use the following recurrence relations. Compute the even part of the continued fraction $g_k(x)$ and the odd part of the continued fraction $g_{k-1}(x)$. These contractions have the same convergents $r_{k,0}, r_{k+1,1}, r_{k+2,2}, \ldots$ and they also have the same form. In this way one can check [2] that:
for $k \geq 1$

$$e_0^{(k)} = 0$$
$$q_1^{(k)} = \frac{f[x_0, \ldots, x_{k+1}]}{f[x_1, \ldots, x_{k+1}]}$$

and for $\ell \geq 1$ and $k \geq 1$

$$e_\ell^{(k)} = \frac{q_\ell^{(k+1)} - q_\ell^{(k)} + e_{\ell-1}^{(k+1)}\left[1 + q_\ell^{(k+1)}(x_0 - x_{k+2\ell-1})\right]}{1 + q_\ell^{(k)}(x_0 - x_{k+2\ell-1})}$$

$$q_{\ell+1}^{(k)} = \frac{e_\ell^{(k+1)} q_\ell^{(k+1)}\left[1 + e_\ell^{(k)}(x_0 - x_{k+2\ell})\right]}{e_\ell^{(k)}\left[1 + q_\ell^{(k+1)}(x_0 - x_{k+2\ell-1})\right] + e_\ell^{(k+1)}(e_\ell^{(k)} - q_\ell^{(k+1)})(x_0 - x_{k+2\ell+1})}$$

These coefficients are usually ordered as in the next table.

Table 3.2.

$$e_0^{(1)}$$
$$q_1^{(1)}$$
$$e_0^{(2)} \qquad\qquad e_1^{(1)}$$
$$q_1^{(2)} \qquad\qquad q_2^{(1)}$$
$$e_0^{(3)} \qquad\qquad e_1^{(2)} \qquad\qquad e_2^{(1)}$$
$$q_1^{(3)} \qquad\qquad q_2^{(2)}$$
$$e_0^{(4)} \qquad\qquad e_1^{(3)} \qquad\qquad e_2^{(2)}$$

Again the superscript denotes a diagonal in the table and the subscript a column. Another qd-like algorithm exists for continued fractions of another form than the one given in (3.6.). Although it is computationally more efficient, it has less interesting properties and so we do not mention it here but refer to [3].

3.5. A generalization of the algorithm of Gragg.

The previous algorithm generalized the qd-algorithm and calculated elements on descending staircases. We can also generalize the algorithm of Gragg and calculate rational interpolants on ascending staircases [3]. To this end we assume normality of the table of rational interpolants.
Consider for $k \geq 1$ the staircase

$$S_k = \{r_{k,0}, r_{k-1,0}, r_{k-1,1}, r_{k-2,1}, \ldots, r_{0,k-1}, r_{0,k}\}$$

and continued fractions of the form

$$f_k(x) = c_0 + \sum_{i=1}^{k} c_i(x - x_0)\ldots(x - x_{i-1}) - \left.\frac{c_k(x - x_0)\ldots(x - x_{k-1})}{1}\right|$$

$$- \left.\frac{f_1^{(k)}}{x - x_k}\right| - \left.\frac{s_1^{(k)}}{1}\right| - \left.\frac{f_2^{(k)}}{x - x_k}\right| - \ldots - \left.\frac{f_k^{(k)}}{x - x_k}\right| \qquad (3.8.)$$

Similar to theorem 3.12. one can prove that there exist coefficients $f_i^{(k)} \neq 0$ and $s_i^{(k)} \neq 0$ such that the successive convergents of f_k are the elements of S_k, as soon as three consecutive elements of S_k are different from each other.

Making use of the relations existing between neighbouring staircases S_k and S_{k+1}, we get the following recurrence relations:

for $k \geq 1$

$$s_0^{(k-1)} = 0$$
$$f_k^{(k-1)} = 0$$
$$f_1^{(k)} = \frac{c_{k-1}}{c_k}$$
$$s_k^{(k)} = \frac{w_{k-1}}{w_k}$$

where $w_k = (1/f)[x_0, \ldots, x_k]$, and for $k \geq 1$ and $1 \leq \ell \leq k-1$

$$s_\ell^{(k)} = s_{\ell-1}^{(k-1)} + f_\ell^{(k-1)} - f_\ell^{(k)} - (x_k - x_{k-1})$$

$$f_{\ell+1}^{(k)} = f_\ell^{(k-1)} \frac{s_\ell^{(k-1)}}{s_\ell^{(k)}}$$

(3.9.)

The coefficients $f_\ell^{(k)}$ and $s_\ell^{(k)}$ can be arranged in a two-dimensional table.

Table 3.3.

	$f_1^{(0)}$		$f_2^{(1)}$		$\cdots$
$s_0^{(0)}$		$s_1^{(1)}$			
	$f_1^{(1)}$		$f_2^{(2)}$		$\cdots$
$s_0^{(1)}$		$s_1^{(2)}$			
	$f_1^{(2)}$		$f_2^{(3)}$		$\cdots$
$s_0^{(2)}$		$s_1^{(3)}$			
$\vdots$	$\vdots$	$\vdots$	$\vdots$		

Each upward sloping diagonal contains the coefficients which are necessary to construct the continued fraction (3.8.). It is easy to see that the formulas (3.9.) reduce to the corresponding algorithm of Gragg for the calculation of Padé approximants in case all the interpolation points coincide with the origin.

3.6. The generalized ε-algorithm.

Let us again consider two neighbouring staircases S_{m+n} and S_{m+n+1}. Each of them can be represented by a continued fraction of the form (3.8.). The successive convergents of the continued fraction constructed from S_{m+n} can be obtained by means of the forward recurrence relations (1.3.). If we write [4]

$$r_{m,n} = \frac{p_{m,n}}{q_{m,n}}$$

then

$$p_{m-1,n+1} = (x - x_{m+n})p_{m-1,n} - f_{n+1}^{(m+n)} \, p_{m,n}$$

$$q_{m-1,n+1} = (x - x_{m+n})q_{m-1,n} - f_{n+1}^{(m+n)} \, q_{m,n}$$

(3.10.)

and

$$p_{m-1,n} = p_{m,n} - s_n^{(m+n)} \, p_{m,n-1}$$

$$q_{m-1,n} = q_{m,n} - s_n^{(m+n)} \, q_{m,n-1}$$

(3.11.)

Consequently, using (3.10.),

$$r_{m-1,n} - r_{m,n} = \frac{p_{m-1,n+1} + f_{n+1}^{(m+n)} p_{m,n}}{q_{m-1,n+1} + f_{n+1}^{(m+n)} q_{m,n}} - \frac{p_{m,n}}{q_{m,n}}$$

$$= \frac{p_{m-1,n+1} q_{m,n} - p_{m,n} q_{m-1,n+1}}{q_{m,n}(q_{m-1,n+1} + f_{n+1}^{(m+n)} q_{m,n})}$$

Using (3.10.) and (3.11.) we get

$$r_{m,n} - r_{m,n-1} = \frac{p_{m,n}}{q_{m,n}} - \frac{p_{m,n}(x - x_{m+n} - f_{n+1}^{(m+n)}) - p_{m-1,n+1}}{q_{m,n}(x - x_{m+n} - f_{n+1}^{(m+n)}) - q_{m-1,n+1}}$$

$$= \frac{p_{m-1,n+1} q_{m,n} - p_{m,n} q_{m-1,n+1}}{q_{m,n}[q_{m,n}(x - x_{m+n} - f_{n+1}^{(m+n)}) - q_{m-1,n+1}]}$$

Combining these two relations, we obtain

$$\left(r_{m-1,n} - r_{m,n}\right)^{-1} + \left(r_{m,n} - r_{m,n-1}\right)^{-1} = \frac{(x - x_{m+n})\dfrac{q_{m,n}}{q_{m-1,n+1}}}{r_{m-1,n+1} - r_{m,n}}$$

Performing analogous operations on S_{m+n+1} we obtain

$$\left(r_{m,n+1} - r_{m,n}\right)^{-1} + \left(r_{m,n} - r_{m+1,n}\right)^{-1} = \frac{(x - x_{m+n+1})\dfrac{q_{m,n}}{q_{m-1,n+1}}}{r_{m-1,n+1} - r_{m,n}}$$

From the last two expressions we easily deduce

$$(x - x_{m+n})^{-1}(r_{m-1,n} - r_{m,n})^{-1} + (x - x_{m+n+1})^{-1}(r_{m+1,n} - r_{m,n})^{-1} =$$

$$(x - x_{m+n})^{-1}(r_{m,n-1} - r_{m,n})^{-1} + (x - x_{m+n+1})^{-1}(r_{m,n+1} - r_{m,n})^{-1}$$

Using this result it is possible to set up the following generalized ε- algorithm [4], in the same way as the ε-algorithm for Padé approximants was constructed from the star identity (2.8.) :

$$\epsilon_{-1}^{(m)} = 0 \quad m = 0, 1, \ldots$$

$$\epsilon_{2n}^{(-n-1)} = 0 \quad n = 0, 1, \ldots$$

$$\epsilon_0^{(m)} = r_{m,0}(x) \quad m = 0, 1, \ldots$$

$$\epsilon_{n+1}^{(m)} = \epsilon_{n-1}^{(m+1)} + \frac{1}{(x - x_{m+n+1})(\epsilon_n^{(m+1)} - \epsilon_n^{(m)})} \qquad \begin{array}{l} m = -\lfloor \frac{n}{2} \rfloor - 1, -\lfloor \frac{n}{2} \rfloor, \ldots \\ n = 0, 1, \ldots \end{array}$$

with

$$\epsilon_{2n}^{(m-n)} = r_{m,n}(x)$$

3.7. Stoer's recursive method.

The use of recursive methods is especially interesting when one needs the function value of an interpolant and not the interpolant itself. Several recursive algorithms were constructed for the rational interpolation problem, one of which is the generalized ϵ-algorithm. Other algorithms can be found in [10, 16, 22]. We shall restrict ourselves here to the presentation of the algorithm described by Stoer. Let

$$p_{m,n}^{(j)}(x) = \sum_{i=0}^{m} a_i\, x^i$$

and

$$q_{m,n}^{(j)}(x) = \sum_{i=0}^{n} b_i\, x^i$$

be defined by

$$(fq_{m,n}^{(j)} - p_{m,n}^{(j)})(x_i) = 0 \quad \text{for} \quad i = j, \ldots, j+m+n$$

in other words, they solve the interpolation problem (3.2.) starting at x_j, and let $a_{m,n}^{(j)}$ and $b_{m,n}^{(j)}$ indicate the coefficients of degree m and n in the polynomials $p_{m,n}^{(j)}$ and $q_{m,n}^{(j)}$ respectively. The following relations describe the successive calculation of the rational interpolants lying on the main descending staircase

$$\left\{ \frac{p_{0,0}^{(j)}}{q_{0,0}^{(j)}},\ \frac{p_{1,0}^{(j)}}{q_{1,0}^{(j)}},\ \frac{p_{1,1}^{(j)}}{q_{1,1}^{(j)}},\ \frac{p_{2,1}^{(j)}}{q_{2,1}^{(j)}}, \ldots \right\}$$

Theorem 3.13.

$$p_{n,n}^{(j)}(x) = (x - x_j)a_{n,n-1}^{(j)}\ p_{n,n-1}^{(j+1)}(x) - (x - x_{j+2n})\, a_{n,n-1}^{(j+1)}\ p_{n,n-1}^{(j)}(x)$$
$$q_{n,n}^{(j)}(x) = (x - x_j)a_{n,n-1}^{(j)}\ q_{n,n-1}^{(j+1)}(x) - (x - x_{j+2n})\, a_{n,n-1}^{(j+1)}\ q_{n,n-1}^{(j)}(x)$$

and
$$\hspace{10cm}(3.12.)$$

$$p_{n+1,n}^{(j)}(x) = (x - x_j)b_{n,n}^{(j)}\ p_{n,n}^{(j+1)}(x) - (x - x_{j+2n+1})\ b_{n,n}^{(j+1)}\ p_{n,n}^{(j)}(x)$$
$$q_{n+1,n}^{(j)}(x) = (x - x_j)b_{n,n}^{(j)}\ q_{n,n}^{(j+1)}(x) - (x - x_{j+2n+1})\ b_{n,n}^{(j+1)}\ q_{n,n}^{(j)}(x)$$

with

$$p_{0,0}^{(j)}(x) = f_j$$

and

$$q_{0,0}^{(j)}(x) = 1$$

Proof

We will perform the proof only for the first set of relations, because the second part is completely analogous. In case one wants to proceed from $p_{n,n-1}^{(j)}$ and $p_{n,n-1}^{(j+1)}$ to $p_{n,n}^{(j)}$ the degree of the numerator may not be raised. The coefficient of the term of degree $n+1$ in the right hand side of (3.12.) is indeed

$$a_{n,n-1}^{(j)} \, a_{n,n-1}^{(j+1)} - a_{n,n-1}^{(j+1)} \, a_{n,n-1}^{(j)} = 0$$

To check the interpolation conditions in x_i for $i = j, \ldots, j+2n$ we divide the set of interpolation points into three subsets:

(a) $(fq_{n,n}^{(j)} - p_{n,n}^{(j)})(x_j) = -(x_j - x_{j+2n}) \, a_{n,n-1}^{(j+1)}(fq_{n,n-1}^{(j)} - p_{n,n-1}^{(j)})(x_j) = 0$

(b) $(fq_{n,n}^{(j)} - p_{n,n}^{(j)})(x_i) = 0$ for $i = j+1, \ldots, j+2n-1$ since

$$(fq_{n,n-1}^{(j+1)} - p_{n,n-1}^{(j+1)})(x_i) = 0$$

and

$$(fq_{n,n-1}^{(j)} - p_{n,n-1}^{(j)})(x_i) = 0$$

(c) $(fq_{n,n}^{(j)} - p_{n,n}^{(j)})(x_{j+2n}) = (x_{j+2n} - x_j) \, a_{n,n-1}^{(j)}(fq_{n,n-1}^{(j+1)} - p_{n,n-1}^{(j+1)})(x_{j+2n}) = 0$ ∎

Again these relations can easily be adapted for the calculation of rational interpolants on other descending staircases. To calculate the interpolants in

$$\left\{ \frac{p_{k,0}^{(j)}}{q_{k,0}^{(j)}}, \frac{p_{k+1,0}^{(j)}}{q_{k+1,0}^{(j)}}, \frac{p_{k+1,1}^{(j)}}{q_{k+1,1}^{(j)}}, \ldots \right\}$$

one starts with

$$p_{k,0}^{(j)} = c_0 + \sum_{i=1}^{k} c_i(x - x_j)\ldots(x - x_{j+i-1})$$

$$q_{k,0}^{(j)} = 1$$

where the c_i are divided differences of f. To calculate the interpolants in

$$\left\{ \frac{p_{0,k}^{(j)}}{q_{0,k}^{(j)}}, \frac{p_{1,k}^{(j)}}{q_{1,k}^{(j)}}, \frac{p_{1,k+1}^{(j)}}{q_{1,k+1}^{(j)}}, \ldots \right\}$$

one starts with

$$p_{0,k}^{(j)} = 1$$

$$q_{0,k}^{(j)} = w_0 + \sum_{i=1}^{k} w_i(x - x_j)\ldots(x - x_{j+i-1})$$

where the w_i are divided differences of $1/f$.

As for Padé approximants, one can also give explicit determinantal formulas for the numerator and denominator of $r_{m,n}(x)$. We will postpone this representation until the next section.

§4. Rational Hermite interpolation.

4.1. Definition of rational Hermite interpolants.

Let the points $\{x_i\}_{i\in\mathbb{N}}$ be distinct and let the numbers $s_i (i \geq 0)$ belong to $\mathbb{N}$. Assume that the derivatives $f^{(\ell)}(x_i)$ of the function f evaluated at the point x_i are given for $\ell = 0,\ldots,s_i - 1$. Consider fixed integers j, k, m and n with

$$1 \leq k \leq s_{j+1}$$

$$m + n + 1 = \sum_{i=0}^{j} s_i + k$$

The **rational Hermite interpolation problem** of order (m,n) for f consists in finding polynomials

$$p(x) = \sum_{i=0}^{m} a_i \, x^i$$

and

$$q(x) = \sum_{i=0}^{n} b_i \, x^i$$

with p/q irreducible and satisfying

$$\begin{cases} f^{(\ell)}(x_i) = \left(\dfrac{p}{q}\right)^{(\ell)}(x_i) & \text{for } \ell = 0,\ldots,s_i - 1 \text{ with } i = 0,\ldots,j \\[2mm] f^{(\ell)}(x_{j+1}) = \left(\dfrac{p}{q}\right)^{(\ell)}(x_{j+1}) & \text{for } \ell = 0,\ldots,k - 1 \end{cases}$$

$$(3.13.)$$

In this interpolation problem s_i interpolation points coincide with x_i, so s_i interpolation conditions must be fulfilled in x_i. Therefore this type of interpolation problem is also often referred to as the osculatory rational interpolation problem [21]. In case $s_i = 1$ for all $i \geq 0$ then the problem is identical to the rational interpolation problem defined at the beginning of this chapter. In case all the interpolation conditions must be satisfied in one single point x_0 then the osculatory rational interpolation problem is identical to the Padé approximation problem defined in the previous chapter.

Instead of considering problem (3.13.) we can look at the linear system of equations

$$\begin{cases} (fq-p)^{(\ell)}(x_i) = 0 & \text{for } \ell = 0, \ldots, s_i - 1 \text{ with } i = 0, \ldots, j \\ (fq-p)^{(\ell)}(x_{j+1}) = 0 & \text{for } \ell = 0, \ldots, k-1 \end{cases} \qquad (3.14.)$$

and this related problem always has a nontrivial solution for $p(x)$ and $q(x)$, since it is a homogeneous system of $m+n+1$ equations in $m+n+2$ unknowns. Again distinct solutions have the same irreducible form p_0/q_0 and we shall call

$$r_{m,n} = \frac{p_0}{q_0}$$

where q_0 is normalized such that $q_0(x_0) = 1$, the **rational Hermite interpolant** of order (m, n) for f.

The rational Hermite interpolation problem can be reformulated as a Newton-Padé approximation problem. We introduce the following notations:

$$y_\ell = x_0 \quad \text{for} \quad \ell = 0, \ldots, s_0 - 1$$
$$y_{d(i)+\ell} = x_i \quad \text{for} \quad \ell = 0, \ldots, s_i - 1 \quad \text{with} \quad d(i) = s_0 + s_1 + \ldots + s_{i-1}(i \geq 1)$$
$$c_{ij} = 0 \quad \text{for} \quad i > j$$
$$c_{ij} = f[y_i, \ldots, y_j] \quad \text{for} \quad i \leq j$$

with possible coalescence of points in the divided difference $f[y_i, \ldots, y_j]$. If we put

$$B_j(x) = \prod_{\ell=1}^{j} (x - y_{\ell-1})$$

with

$$B_0(x) = 1$$

then formally

$$f(x) = \sum_{i=0}^{\infty} c_{0i} \, B_i(x)$$

This series is called the **Newton series** for f.
Problem (3.14.) is then equivalent with the computation of polynomials

$$p(x) = \sum_{i=0}^{m} a_i \, B_i(x)$$

and

$$q(x) = \sum_{i=0}^{n} b_i \, B_i(x)$$

such that

$$(f \, q - p)(x) = \sum_{i \geq m+n+1} d_i \, B_i(x) \tag{3.15.}$$

Problem (3.15.) is called the **Newton-Padé approximation problem** of order (m, n) for f.

To determine solutions p and q of (3.15.) the divided differences

$$d_i = (f \, q - p)[y_0, \ldots, y_i] \quad \text{for} \quad i = 0, \ldots, m+n$$

must be calculated and put equal to zero. The following lemma, which is a generalization of the Leibniz rule for differentiating a product of functions, is a useful tool.

Lemma 3.1.

$$(f \, q)[y_0, \ldots, y_i] = \sum_{\ell=0}^{i} \, f[y_0, \ldots, y_\ell] \, q[y_\ell, \ldots, y_i]$$

For the proof we refer to [19].

Using lemma 3.1. it is now possible to write down the linear systems of equations that must be satisfied by the coefficients a_i and b_i in p and q :

$$\begin{cases} c_{00} \, b_0 = a_0 \\ c_{01} \, b_0 + c_{11} \, b_1 = a_1 \\ \quad \vdots \\ c_{0m} b_0 + c_{1m} b_1 + \ldots + c_{nm} b_n = a_m \end{cases} \tag{3.16a.}$$

$$\begin{cases} c_{0,m+1} \, b_0 + \ldots + c_{n,m+1} \, b_n = 0 \\ \quad \vdots \\ c_{0,m+n} \, b_0 + \ldots + c_{n,m+n} \, b_n = 0 \end{cases} \tag{3.16b.}$$

Since the problems (3.14.) and (3.15.) are equivalent, the rational function $r_{m,n}$ can as well be called the **Newton-Padé approximant** of order (m, n) to f. In the same way as for the rational interpolation problem the following theorem can be proved.

Theorem 3.14.

The rational Hermite interpolation problem (3.13.) has a solution if and only if the rational Hermite interpolant $r_{m,n} = p_0/q_0$ satisfies (3.14.).

4.2. The table of rational Hermite interpolants.

Once again we will order the interpolants $r_{m,n}$ in a table with double entry:

$$
\begin{array}{cccc}
r_{0,0} & r_{0,1} & r_{0,2} & \cdots \\[1ex]
r_{1,0} & r_{1,1} & r_{1,2} & \cdots \\[1ex]
r_{2,0} & r_{2,1} & r_{2,2} & \cdots \\[1ex]
r_{3,0} & \vdots & \vdots & \\[1ex]
\vdots & & &
\end{array}
$$

For a detailed study of the structure of the rational Hermite interpolation table we refer to [3]. We will only summarize some results. They are based on the following property.

Theorem 3.15.

If the rank of the linear system (3.16b.) is $n - t$ then (up to a normalization) a unique solution $\bar{p}(x)$ and $\bar{q}(x)$ of (3.16.) exists with

$$\partial \bar{p} \leq m - t$$
$$\partial \bar{q} \leq n - t$$

where at least one of the upper bounds is attained. Every other solution $p(x)$ and $q(x)$ of (3.16.) can be written in the form

$$p(x) = \bar{p}(x)\, s(x)$$
$$q(x) = \bar{q}(x)\, s(x)$$

where $\partial s \leq t$.

Proof

Since the rank of the linear system (3.16b.) is $n - t$, solutions p_1, q_1 and p_2, q_2 of (3.16.) can be constructed such that

$$\partial p_1 \leq m$$
$$\partial q_1 \leq n - t$$

and

$$\partial p_2 \leq m - t$$
$$\partial q_2 \leq n$$

by choosing the free parameters of the homogeneous system in an appropriate way.

Then one can prove, just as in theorem 3.1., that

$$p_1 q_2 = p_2 q_1$$

Since

$$\partial(p_2 q_1) \leq m + n - 2t$$

we must have either $\partial p_1 \leq m - t$ or $\partial q_2 \leq n - t$. Hence there exists a solution $\bar{p}, \bar{q}$ of (3.16.) with $\partial \bar{p} \leq m - t$ and $\partial \bar{q} \leq n - t$ and it is easy to see that it is unique (up to a normalization).

Now the polynomials

$$p_{(i)}(x) = B_i(x)\bar{p}(x)$$

and

$$q_{(i)}(x) = B_i(x)\bar{q}(x)$$

with $0 \leq i \leq t$ do also solve the Newton-Padé approximation problem of order (m, n) because already

$$(f\bar{q} - \bar{p})(x) = \sum_{i \geq m+n+1} d_i B_i(x)$$

What's more, they are a linearly independent set of solutions and hence span the solution space. Consequently, all other solutions p and q of 3.16. are a polynomial multiple of $\bar{p}$ and $\bar{q}$.

Now if both $\partial \bar{p} < m - t$ and $\partial \bar{q} < n - t$ we would have more than $t + 1$ linearly independent solutions and thus the dimension of the solution space would be greater than $t + 1$ which implies that the rank of the system (3.16b.) would be less than $n - t$ and this is a contradiction. ∎

Before describing the shape of sets in the table of rational Hermite interpolants that contain equal elements, it is important to emphasize that the structure of the table can only be studied if the ordering of the interpolation points $\{x_i\}_{i \in \mathbb{N}}$ remains fixed once it is chosen.

Since the polynomials $\bar{p}$ and $\bar{q}$ constructed in the previous theorem have the property that their degrees cannot be lowered simultaneously anymore unless some interpolation conditions are lost, we shall call them a **minimal solution**. This does not imply that $\bar{p}/\bar{q}$ is irreducible. However we still have

$$p_0 \bar{q} = \bar{p} q_0$$

where $p_0/q_0 = r_{m,n}$.

Theorem 3.16.

Let $\bar{p}(x)$ and $\bar{q}(x)$ be the minimal solution of the Newton-Padé approximation problem of order (m, n) for f and let the rank of the linear system (3.16b.) be $n - t$.

a) If $\partial \bar{p} = m - t - t_1$ then all the rational Hermite interpolants lying in the triangle with corner elements $r_{m-t-t_1,n-t}$, $r_{m-t-t_1,n+t+t_1}$ and $r_{m+t,n-t}$ are equal to $r_{m-t-t_1,n-t}$.

b) If $\partial \bar{q} = n - t - t_2$ then all the rational Hermite interpolants lying in the triangle with corner elements $r_{m-t,n-t-t_2}$, $r_{m-t,n+t}$ and $r_{m+t+t_2,n-t-t_2}$ are equal to $r_{m-t,n-t-t_2}$.

c) If

$$(f\bar{q} - \bar{p})(x) = \sum_{i \geq m+n-2t+t_3+1} d_i B_i(x)$$

with $d_{m+n-2t+t_3+1} \neq 0$ then all the rational Hermite interpolants lying in the triangle with corner elements $r_{m-t,n-t}$, $r_{m+t+t_3,n-t}$ and $r_{m-t,n+t+t_3}$ are equal to $r_{m-t,n-t}$.

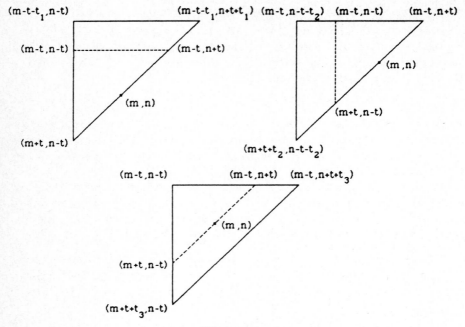

Figure 3.1.

For the proof we refer to [5].

4.3. Determinant representation.

Theorem 3.17.

If the rank of the system of equations (3.16b.) is maximal, then (up to a normalization)

$$r_{m,n} = \frac{p_0}{q_0}$$

is given by

$$p_0(x) = \begin{vmatrix} F_{0,m}(x) & F_{1,m}(x) & \cdots & F_{n,m}(x) \\ c_{0,m+1} & c_{1,m+1} & \cdots & c_{n,m+1} \\ c_{0,m+2} & c_{1,m+2} & \cdots & c_{n,m+2} \\ \vdots & \vdots & \ddots & \vdots \\ c_{0,m+n} & c_{1,m+n} & \cdots & c_{n,m+n} \end{vmatrix}$$

$$q_0(x) = \begin{vmatrix} B_0(x) & B_1(x) & \dots & B_n(x) \\ c_{0,m+1} & c_{1,m+1} & \cdots & c_{n,m+1} \\ c_{0,m+2} & c_{1,m+2} & \cdots & c_{n,m+2} \\ \vdots & \vdots & \ddots & \vdots \\ c_{0,m+n} & c_{1,m+n} & \cdots & c_{n,m+n} \end{vmatrix}$$

where

$$F_{i,j}(x) = \sum_{\ell=i}^{j} c_{i\ell}\, B_\ell(x) \;\; \text{if} \;\; i \le j$$

and

$$F_{i,j}(x) = 0 \;\; \text{if} \;\; i > j$$

The proof is left as an exercise (see problem (5)).
Again one can see that in case all the interpolation points coincide with one single point, these determinant formulas reduce to the ones given in the chapter on Padé approximants since the divided differences reduce to Taylor coefficients.

4.4. Continued fraction representation.

If one considers staircases

$$T_k = \{r_{k,0}, r_{k+1,0}, r_{k+1,1}, r_{k+2,1}, \ldots\}$$

in the table of rational Hermite interpolants, one can again construct continued fractions of which the successive convergents equal the elements of T_k. It is easy to see that these continued fractions are of the form

$$d_0 + d_1\, B_1(x) + \ldots + d_k\, B_k(x) + \left|\frac{d_{k+1}\, B_{k+1}(x)}{1}\right|$$

$$+ \left|\frac{d_{k+2}(x - y_{k+1})}{1}\right| + \left|\frac{d_{k+3}(x - y_{k+2})}{1}\right| + \ldots$$

The coefficients $d_0, \ldots, d_{k+1}$ are divided differences (with coalescence of points) and the other coefficients can still be obtained using the generalized qd-algorithm. The generalization of Gragg's algorithm and the generalized ϵ-algorithm also remain valid for the calculation of rational Hermite interpolants.

4.5. Thiele's continued fraction expansion.

From theorem 3.9. we write formally

$$f(x) = \varphi_0[x_0] + \frac{x - x_0}{\left|\varphi_1[x_0, x_1]\right.} + \frac{x - x_1}{\left|\varphi_2[x_0, x_1, x_2]\right.} + \ldots$$

$$+ \frac{x - x_{i-1}}{\left|\varphi_i[x_0, \ldots, x_i]\right.} + \ldots \qquad (3.17.)$$

We consider now the limiting case

$$x_i \to x_0 \quad \text{for} \quad i \geq 1$$

$$\varphi_j(x) = \lim_{\substack{x_i \to x \\ i = 0, \ldots, j}} \varphi_j[x_0, \ldots, x_j]$$

Then (3.17.) becomes

$$f(x) = \varphi_0(x_0) + \frac{x - x_0}{\left|\varphi_1(x_0)\right.} + \frac{x - x_0}{\left|\varphi_2(x_0)\right.} + \ldots + \frac{x - x_0}{\left|\varphi_i(x_0)\right.} + \ldots \qquad (3.18.)$$

This is a **continued fraction expansion** of f around x_0. Formula (3.18.) is obtained from (3.17.) in the same way as we set up a Taylor series development from Newton's interpolation formula. Since (3.18.) is formal, one has to check for which values of x the righthand side really converges to $f(x)$. We can calculate the $\varphi_j(x)$ using Thiele's method [17] :

$$\varphi_0(x) = f(x)$$

$$\varphi_1(x) = \lim_{x_0, x_1 \to x} \varphi_1[x_0, x_1]$$

$$= \lim_{x_1 \to x} \frac{x - x_1}{\varphi_0[x] - \varphi_0[x_1]} = \frac{1}{f'(x)}$$

and with

$$\rho_j(x) = \lim_{\substack{x_i \to x \\ i = 0, \ldots, j}} \rho_j[x_0, \ldots, x_j]$$

we have

$$\varphi_j(x) = \lim_{\substack{x_i \to x \\ i = 0, \ldots, j}} \frac{x_j - x_{j-1}}{\rho_{j-1}[x_0, \ldots, x_{j-2}, x_j] - \rho_{j-1}[x_0, \ldots, x_{j-2}, x_{j-1}]}$$

$$= \lim_{x_j \to x} \frac{x - x_j}{\rho_{j-1}[x, \ldots, x] - \rho_{j-1}[x, \ldots, x, x_j]}$$

$$= \left[\frac{\partial \rho_{j-1}[x, \ldots, x, y]}{\partial y} \Big|_{y=x} \right]^{-1}$$

Now $\rho_{j-1}[x_0, \ldots, x_{j-1}]$ does not depend upon the ordering of the points $x_0, \ldots, x_{j-1}$. So

$$\frac{d\rho_{j-1}[x, \ldots, x]}{dx} = j \cdot \frac{\partial \rho_{j-1}[x, \ldots, x, y]}{\partial y} \Big|_{y=x}$$

Consequently

$$\varphi_j(x) = j \left(\frac{d\rho_{j-1}(x)}{dx} \right)^{-1} \tag{3.19a.}$$

To calculate

$$\frac{d\rho_{j-1}(x)}{dx}$$

one uses the relationship

$$\rho_j(x) = \rho_{j-2}(x) + \varphi_j(x) \tag{3.19b.}$$

We apply the formulas (3.19.) to construct a continued fraction expansion of $f(x) = e^x$ around the origin:

$\rho_0(x) = e^x$	$\varphi_1(x) = e^{-x}$	$\varphi_1(x_0) = 1$
$\rho_1(x) = e^{-x}$	$\varphi_2(x) = -2e^x$	$\varphi_2(x_0) = -2$
$\rho_2(x) = -e^x$	$\varphi_3(x) = -3e^{-x}$	$\varphi_3(x_0) = -3$
$\rho_3(x) = -2e^{-x}$	$\varphi_4(x) = 2e^x$	$\varphi_4(x_0) = 2$
$\rho_4(x) = e^x$	$\varphi_5(x) = 5e^{-x}$	$\varphi_5(x_0) = 5$
$\vdots$	$\vdots$	$\vdots$

So we get

$$e^x = 1 + \frac{x}{|1} + \frac{x}{|-2} + \frac{x}{|-3} + \frac{x}{|2} + \frac{x}{|5} + \ldots$$

§5. Convergence of rational Hermite interpolants.

The theorems of chapter I can be used to investigate the convergence of interpolating continued fractions. We shall now mention some results for the convergence of columns in the table of rational Hermite interpolants.

The first theorem deals with the first column $\{r_{0,0}, r_{1,0}, r_{2,0}, \ldots\}$, in other words it is a convergence theorem for interpolating polynomials.

For given complex points $\{z_0, \ldots, z_j\}$ we define the **lemniscate**

$$B(z_0, \ldots, z_j, r) = \{z \in \mathbb{C} \mid |(z - z_0)(z - z_1) \ldots (z - z_j)| = r\}$$

Broadly speaking, the convergence of an arbitrary series of interpolation does not depend on the entire sequence of interpolation points y_i (as defined in the Newton-Padé approximation problem) but merely on its asymptotic character, as can be seen in the next theorem.

Theorem 3.18.

Let the sequence of interpolation points $\{y_0, y_1, y_2, \ldots\}$ be asymptotic to the sequence

$$\{z_0, z_1, \ldots, z_j, z_0, z_1, \ldots, z_j, z_0, z_1, \ldots, z_j, \ldots\}$$

in the sense that

$$\lim_{k \to \infty} y_{k(j+1)+i} = z_i$$

for $i = 0, \ldots, j$.

If the function $f(x)$ is analytic throughout the interior of the lemniscate $B(z_0, \ldots, z_j, r)$ then the $r_{m,0}$ converge to f on the interior of $B(z_0, \ldots, z_j, r)$. The convergence is uniform on every closed and bounded subset interior to $B(z_0, \ldots, z_j, r)$.

For the proof we refer to [20 p. 61] and [9 pp. 90-91].

Let us now turn to the case of a meromorphic function f with poles $w_1, \ldots, w_n$ (counted with their multiplicity). For the rational Hermite interpolant of order (m, n) we write

$$r_{m,n} = \frac{p_{m,n}}{q_{m,n}}$$

and for the minimal solution of the Newton-Padé approximation problem of order (m, n) we write $\bar{p}_{m,n}(x)$ and $\bar{q}_{m,n}(x)$.

Let the table of minimal solutions for the Newton-Padé approximation problem be normal. According to [6] we then have $\partial \bar{q}_{m,n} = n$. Let $w_i^{(m)}$ $(i = 1, \ldots, n)$ be the zeros of $q_{m,n}$ for $m = 0, 1, 2, \ldots$ and let $\rho_i = |(w_i - z_0)(w_i - z_1)\ldots(w_i - z_j)|$ with $0 < \rho_1 \leq \rho_2 \leq \ldots \leq \rho_n \leq \alpha r < r$ for a positive constant α.

Theorem 3.19.

If the sequence of interpolation points $\{y_0, y_1, y_2, \ldots\}$ is asymptotic to the sequence $\{z_0, z_1, \ldots, z_j, z_0, z_1, \ldots, z_j, z_0, z_1, \ldots, z_j, \ldots\}$, if f is meromorphic in the interior of $B(z_0, \ldots, z_j, r)$ with poles $w_1, \ldots, w_n$ counted with their multiplicity and if the table of minimal solutions for the Newton-Padé approximation problem is normal, then

$$w_i^{(m)} = w_i + 0(\alpha^m) \qquad i = 1, \ldots, n$$

and

$$\lim_{m \to \infty} r_{m,n}(x) = f(x)$$

uniformly in every closed and bounded subset of the interior of $B(z_0, \ldots, z_j, r)$ not containing the points $w_1, \ldots, w_n$.

The proof is given in [4].

§6. Multivariate rational interpolants.

We have seen that univariate rational interpolants can be obtained in various equivalent ways: one can calculate the explicit solution of the system of interpolatory conditions, start a recursive algorithm, calculate the convergent of a continued fraction or solve the Newton-Padé approximation problem. We will generalize the last two techniques for multivariate functions. These generalizations are written down for the case of two variables, because the situation with more than two variables is only notationally more difficult. More details can be found in [7] and [8].

6.1. Interpolating branched continued fractions.

Given two sequences $\{x_0, x_1, x_2, \ldots\}$ and $\{y_0, y_1, y_2, \ldots\}$ of distinct points we will interpolate the bivariate function $f(x, y)$ at the points in $\{x_0, x_1, x_2, \ldots\} \times \{y_0, y_1, y_2, \ldots\}$. To this end we use branched continued fractions symmetric in the variables x and y and we define **bivariate inverse differences** as follows:

$$\varphi_{0,0}[x_0][y_0] = f(x_0, y_0)$$

$$\varphi_{k,0}[x_0, \ldots, x_k][y_0]$$
$$= \frac{x_k - x_{k-1}}{\varphi_{k-1,0}[x_0, \ldots, x_{k-2}, x_k][y_0] - \varphi_{k-1,0}[x_0, \ldots, x_{k-2}, x_{k-1}][y_0]}$$

$$\varphi_{0,k}[x_0][y_0, \ldots, y_k]$$
$$= \frac{y_k - y_{k-1}}{\varphi_{0,k-1}[x_0][y_0, \ldots, y_{k-2}, y_k] - \varphi_{0,k-1}[x_0][y_0, \ldots, y_{k-2}, y_{k-1}]}$$

$$\varphi_{j,j}[x_0, \ldots, x_j][y_0, \ldots, y_j]$$
$$= \frac{(x_j - x_{j-1})(y_j - y_{j-1})}{\begin{array}{l}\varphi_{j-1,j-1}[x_0, \ldots, x_{j-2}, x_j][y_0, \ldots, y_{j-2}, y_j] - \varphi_{j-1,j-1}[x_0, \ldots, x_{j-2}, x_{j-1}][y_0, \ldots, y_{j-2}, y_j] \\ -\varphi_{j-1,j-1}[x_0, \ldots, x_{j-2}, x_j][y_0, \ldots, y_{j-2}, y_{j-1}] + \varphi_{j-1,j-1}[x_0, \ldots, x_{j-2}, x_{j-1}][y_0, \ldots, y_{j-2}, y_{j-1}]\end{array}}$$

and for $k > j$
$$\varphi_{k,j}[x_0, \ldots, x_k][y_0, \ldots, y_j]$$
$$= \frac{x_k - x_{k-1}}{\varphi_{k-1,j}[x_0, \ldots, x_{k-2}, x_k][y_0, \ldots, y_j] - \varphi_{k-1,j}[x_0, \ldots, x_{k-2}, x_{k-1}][y_0, \ldots, y_j]}$$

$$\varphi_{j,k}[x_0, \ldots, x_j][y_0, \ldots, y_k]$$
$$= \frac{y_k - y_{k-1}}{\varphi_{j,k-1}[x_0, \ldots, x_j][y_0, \ldots, y_{k-2}, y_k] - \varphi_{j,k-1}[x_0, \ldots, x_j][y_0, \ldots, y_{k-2}, y_{k-1}]}$$

Theorem 3.20.

$$f(x,y) = B_0(x,y) + \sum_{j=1}^{\infty} \frac{(x - x_{j-1})(y - y_{j-1})}{B_j(x,y)} \bigg| \qquad (3.20.)$$

with

$$B_j(x,y) = \varphi_{j,j}[x_0, \ldots, x_j][y_0, \ldots, y_j]$$

$$+ \sum_{k=j+1}^{\infty} \frac{x - x_{k-1}}{\varphi_{k,j}[x_0, \ldots, x_k][y_0, \ldots, y_j]} \bigg|$$

$$+ \sum_{k=j+1}^{\infty} \frac{y - y_{k-1}}{\varphi_{j,k}[x_0, \ldots, x_j][y_0, \ldots, y_k]} \bigg|$$

Proof

From theorem 3.9. we know that

$$f(x,y) = \varphi[x][y]$$
$$= \varphi_{0,0}[x_0][y] + \frac{x - x_0}{\varphi_{1,0}[x_0, x][y]}$$
$$= \varphi_{0,0}[x_0][y_0] + \frac{y - y_0}{\varphi_{0,1}[x_0][y_0, y]} + \frac{x - x_0}{\varphi_{1,0}[x_0, x][y]}$$
$$= \varphi_{0,0}[x_0][y_0] + \sum_{k=1}^{\infty} \frac{y - y_{k-1}}{\varphi_{0,k}[x_0][y_0, \ldots, y_k]} \bigg| + \frac{x - x_0}{\varphi_{1,0}[x_0, x][y]}$$

Let us introduce the function $g_0(x,y)$ by

$$\frac{x - x_0}{\varphi_{1,0}[x_0, x][y]} = (x - x_0)g_0(x,y)$$

where

$$g_0(x,y) = \frac{1}{\varphi_{1,0}[x_0, x][y]}$$

By calculating inverse differences $\xi_{j,k}^{(0)}$ for g_0 we obtain

$$g_0(x,y) = \xi_{0,0}^{(0)}[x][y]$$

$$= \xi_{0,0}^{(0)}[x][y_0] + \frac{y-y_0}{\xi_{0,1}^{(0)}[x][y_0,y]}$$

$$= \frac{1}{\varphi_{1,0}[x_0,x][y_0]} + \frac{y-y_0}{\xi_{0,1}^{(0)}[x][y_0,y]}$$

$$= \frac{1}{\varphi_{1,0}[x_0,x_1][y_0] + \dfrac{x-x_1}{\varphi_{2,0}[x_0,x_1,x][y_0]}} + \frac{y-y_0}{\xi_{0,1}^{(0)}[x][y_0,y]}$$

$$= \frac{1}{\varphi_{1,0}[x_0,x_1][y_0] + \sum_{k=2}^{\infty} \left|\dfrac{x-x_{k-1}}{\varphi_{k,0}[x_0,\ldots,x_k][y_0]}\right.} + \frac{y-y_0}{h_0(x,y)}$$

where $h_0(x,y) = \xi_{0,1}^{(0)}[x][y_0,y]$. So already

$$f(x,y) = \varphi_{0,0}[x_0][y_0] + \sum_{k=1}^{\infty} \left|\frac{y-y_{k-1}}{\varphi_{0,k}[x_0][y_0,\ldots,y_k]}\right.$$

$$+ \sum_{k=1}^{\infty} \left|\frac{x-x_{k-1}}{\varphi_{k,0}[x_0,\ldots,x_k][y_0]}\right.$$

$$+ \frac{(x-x_0)(y-y_0)}{h_0(x,y)}$$

By computing inverse differences $\pi_{j,k}^{(0)}$ for h_0 we get

$$h_0(x,y) = \pi_{0,0}^{(0)}[x][y]$$

$$= \pi_{0,0}^{(0)}[x_1][y] + \frac{x-x_1}{\pi_{1,0}^{(0)}[x_1,x][y]}$$

$$= \pi_{0,0}^{(0)}[x_1][y_1] + \frac{y-y_1}{\pi_{0,1}^{(0)}[x_1][y_1,y]} + \frac{x-x_1}{\pi_{1,0}^{(0)}[x_1,x][y]}$$

$$= \pi_{0,0}^{(0)}[x_1][y_1] + \sum_{k=2}^{\infty} \left|\frac{y-y_{k-1}}{\pi_{0,k-1}^{(0)}[x_1][y_1,\ldots,y_k]}\right. + \frac{x-x_1}{\pi_{1,0}^{(0)}[x_1,x][y]}$$

It is easy to see that

$$\pi_{0,0}^{(0)}[x][y] = \varphi_{1,1}[x_0, x][y_0, y]$$

From this we find by induction that

$$\pi_{0,k-1}^{(0)}[x_1][y_1, \ldots, y_k] = \varphi_{1,k}[x_0, x_1][y_0, \ldots, y_k]$$
$$\pi_{1,0}^{(0)}[x_1, x][y] = \varphi_{2,1}[x_0, x_1, x][y_0, y]$$

So we can write

$$h_0(x, y) = \varphi_{1,1}[x_0, x_1][y_0, y_1] + \sum_{k=2}^{\infty} \left. \frac{y - y_{k-1}}{\varphi_{1,k}[x_0, x_1][y_0, \ldots, y_k]} \right|$$
$$+ \frac{x - x_1}{\varphi_{2,1}[x_0, x_1, x][y_0, y]}$$
$$= \varphi_{1,1}[x_0, x_1][y_0, y_1] + \sum_{k=2}^{\infty} \left. \frac{y - y_{k-1}}{\varphi_{1,k}[x_0, x_1][y_0, \ldots, y_k]} \right|$$
$$+ (x - x_1) \ g_1(x, y)$$

where

$$g_1(x, y) = \frac{1}{\varphi_{2,1}[x_0, x_1, x][y_0, y]}$$

If we introduce inverse differences $\xi_{j,k}^{(1)}$ for g_1 we can repeat the whole reasoning which provides us with a function h_1 and inverse differences $\pi_{j,k}^{(1)}$:

$$g_1(x, y) = \xi_{0,0}^{(1)}[x][y] = \xi_{0,0}^{(1)}[x][y_1] + \frac{y - y_1}{\xi_{0,1}^{(1)}[x][y_1, y]}$$

$$= \cfrac{1}{\varphi_{2,1}[x_0, x_1, x_2][y_0, y_1] + \sum_{k=3}^{\infty} \left. \cfrac{x - x_{k-1}}{\varphi_{k,1}[x_0, x_1, \ldots, x_k][y_0, y_1]} \right| }$$
$$+ \frac{y - y_1}{\xi_{0,1}^{(1)}[x][y_1, y]}$$

and

$$h_1(x, y) = \xi_{0,1}^{(1)}[x][y_1, y]$$
$$= \pi_{0,0}^{(1)}[x_2][y_2] + \sum_{k=3}^{\infty} \left. \frac{y - y_{k-1}}{\pi_{0,k-2}^{(1)}[x_2][y_2, \ldots, y_k]} \right| + \frac{x - x_2}{\pi_{1,0}^{(1)}[x_2, x][y]}$$

In the same way as for h_0 we find

$$h_1(x,y) = \varphi_{2,2}[x_0,x_1,x_2][y_0,y_1,y_2] + \sum_{k=3}^{\infty} \left. \frac{y - y_{k-1}}{\varphi_{2,k}[x_0,x_1,x_2][y_0,\ldots,y_k]} \right|$$
$$+ \frac{x - x_2}{\varphi_{3,2}[x_0,x_1,x_2,x][y_0,y_1,y]}$$

Finally we obtain the desired interpolatory continued fraction. ∎

To obtain rational interpolants we are going to consider convergents of the branched continued fraction (3.20.). To indicate which convergent we compute we need a multi-index

$$\overline{n} = (n, m_{0x}, m_{0y}, \ldots, m_{nx}, m_{ny})$$

The $\overline{n}^{th}$ convergent is then given by

$$C_{\overline{n}}(x,y) = \varphi_{0,0}[x_0][y_0] + \sum_{k=1}^{m_{0x}} \left. \frac{x - x_{k-1}}{\varphi_{k,0}[x_0,\ldots,x_k][y_0]} \right|$$
$$+ \sum_{k=1}^{m_{0y}} \left. \frac{y - y_{k-1}}{\varphi_{0,k}[x_0][y_0,\ldots,y_k]} \right|$$
$$+ \sum_{j=1}^{n} \left. \frac{(x - x_{j-1})(y - y_{j-1})}{C^{(j)}_{m_{jx}, m_{jy}}(x,y)} \right|$$

with

$$C^{(j)}_{m_{jx}, m_{jy}}(x,y) = \varphi_{j,j}[x_0,\ldots,x_j][y_0,\ldots,y_j]$$
$$+ \sum_{k=j+1}^{m_{jx}} \left. \frac{x - x_{k-1}}{\varphi_{k,j}[x_0,\ldots,x_k][y_0,\ldots,y_j]} \right|$$
$$+ \sum_{k=j+1}^{m_{jy}} \left. \frac{y - y_{k-1}}{\varphi_{j,k}[x_0,\ldots,x_j][y_0,\ldots,y_k]} \right|$$

For these rational functions the following interpolation property can be proved.

Theorem 3.21.

The convergent $C_{\overline{n}}(x, y)$ of (3.20.) satisfies

$$C_{\overline{n}}(x_{\ell_1}, y_{\ell_2}) = f(x_{\ell_1}, y_{\ell_2})$$

for (ℓ_1, ℓ_2) belonging to

$$I = \bigcup_{j=0}^{n} \Big[\{(j,k)| \forall 0 \le i \le j : j \le k \le m_{iy}\} \bigcup \{(k,j)| \forall 0 \le i \le j : j \le k \le m_{ix}\} \Big]$$

if $C_{\overline{n}}(x_{\ell_1}, y_{\ell_2})$ is defined.

Proof

Let $\ell = \min(\ell_1, \ell_2)$. From theorem 3.20. we know that

$$f(x_{\ell_1}, y_{\ell_2}) = \varphi_{0,0}[x_0][y_0] + \sum_{k=1}^{\ell_1} \left| \frac{x_{\ell_1} - x_{k-1}}{\varphi_{k,0}[x_0, \ldots, x_k][y_0]} \right. + \sum_{k=1}^{\ell_2} \left| \frac{y_{\ell_2} - y_{k-1}}{\varphi_{0,k}[x_0][y_0, \ldots, y_k]} \right.$$
$$+ \sum_{j=1}^{\ell} \left| \frac{(x_{\ell_1} - x_{j-1})(y_{\ell_2} - y_{j-1})}{B_j(x_{\ell_1}, y_{\ell_2})} \right.$$

where

$$B_j(x_{\ell_1}, y_{\ell_2}) = \varphi_{j,j}[x_0, \ldots, x_j][y_0, \ldots, y_j]$$
$$+ \sum_{k=j+1}^{\ell_1} \left| \frac{x_{\ell_1} - x_{k-1}}{\varphi_{k,j}[x_0, \ldots, x_k][y_0, \ldots, y_j]} \right.$$
$$+ \sum_{k=j+1}^{\ell_2} \left| \frac{y_{\ell_2} - y_{k-1}}{\varphi_{j,k}[x_0, \ldots, x_j][y_0, \ldots, y_k]} \right.$$

Now $f(x_{\ell_1}, y_{\ell_2}) = C_{\overline{n}}(x_{\ell_1}, y_{\ell_2})$ if and only if the following conditions are satisfied

$$\ell \le n$$
$$\forall \, 0 \le i \le \ell : \ell_1 \le m_{ix} \text{ and } \ell_2 \le m_{iy}$$

This is precisely guaranteed by saying $(\ell_1, \ell_2) \in I$.   ■

For instance, if $m_{0x} \geq m_{1x} \geq \ldots \geq m_{nx}$ and $m_{0y} \geq m_{1y} \geq \ldots \geq m_{ny}$ the boundary of the set I is given by $\bar{n} = (n, m_{0x}, m_{0y}, \ldots, m_{nx}, m_{ny})$, as we can tell from the next picture which is drawn for $n = 2$.

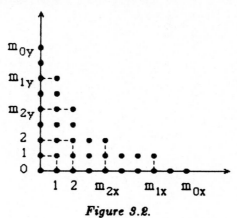

Figure 3.2.

We illustrate this technique with a simple numerical example.

Let the following data be given: $x_i = i$ for $i = 0, 1, 2, \ldots$ and $y_j = j$ for $j = 0, 1, 2, \ldots$ with $f(x_i, y_j) = (i + j)^2$.

Take $\bar{n} = (1, 2, 2, 1, 1)$. Then we have to compute

$$\varphi_{0,0}[x_0][y_0]$$
$$\varphi_{1,0}[x_0, x_1][y_0] \qquad \varphi_{0,1}[x_0][y_0, y_1]$$
$$\varphi_{2,0}[x_0, x_1, x_2][y_0] \qquad \varphi_{0,2}[x_0][y_0, y_1, y_2]$$
$$\varphi_{1,1}[x_0, x_1][y_0, y_1]$$

The resulting convergent is

$$C_{\bar{n}} = \frac{x}{1 - \dfrac{x - 1}{2}} + \frac{y}{1 - \dfrac{y - 1}{2}} + 2xy$$

It interpolates f in the points $(x_0, y_0), (x_1, y_0), (x_2, y_0), (x_0, y_1), (x_0, y_2), (x_1, y_1)$.

6.2. General order Newton-Padé approximants.

Consider two sequences of real points $\{x_i\}_{i \in \mathbb{N}}$ and $\{y_j\}_{j \in \mathbb{N}}$ where coalescent points get consecutive numbers. For a bivariate function $f(x, y)$ we define the following divided differences

$$f[x_0][y_0] = f(x_0, y_0)$$
$$f[x_0][y_0, \ldots, y_k] =$$
$$\frac{f[x_0][y_1, \ldots, y_k] - f[x_0][y_0, \ldots, y_{k-1}]}{y_k - y_0}$$
$$f[x_0, \ldots, x_k][y_0] =$$
$$\frac{f[x_1, \ldots, x_k][y_0] - f[x_0, \ldots, x_{k-1}][y_0]}{x_k - x_0}$$
$$f[x_0, \ldots, x_k][y_0, \ldots, y_\ell] =$$
$$\frac{f[x_0, \ldots, x_k][y_1, \ldots, y_\ell] - f[x_0, \ldots, x_k][y_0, \ldots, y_{\ell-1}]}{y_\ell - y_0} \tag{3.21a.}$$

or equivalently

$$f[x_0, \ldots, x_k][y_0, \ldots, y_\ell] =$$
$$\frac{f[x_1, \ldots, x_k][y_0, \ldots, y_\ell] - f[x_0, \ldots, x_{k-1}][y_0, \ldots, y_\ell]}{x_k - x_0} \tag{3.21b.}$$

One can easily prove that (3.21a.) and (3.21b.) give the same result. When the interpolation points $x_i, \ldots, x_{i+r_i-1}$ and $y_j, \ldots, y_{j+s_j-1}$ coincide, then one must bear in mind that

$$f[x_i, \ldots, x_{i+r_i-1}][y_\ell] = \frac{1}{r_i!} \frac{\partial^{r_i-1} f}{\partial x^{r_i-1}}\Big|_{(x_i, y_\ell)}$$

$$f[x_k][y_j, \ldots, y_{j+s_j-1}] = \frac{1}{s_j!} \frac{\partial^{s_j-1} f}{\partial y^{s_j-1}}\Big|_{(x_k, y_j)}$$

$$f[x_i, \ldots, x_{i+r_i-1}][y_j, \ldots, y_{j+s_j-1}] = \frac{1}{r_i!} \frac{1}{s_j!} \frac{\partial^{r_i+s_j-2} f}{\partial x^{r_i-1} \partial y^{s_j-1}}\Big|_{(x_i, y_j)}$$

We consider the following set of basis functions for the real-valued polynomials in two variables:

$$B_{ij}(x, y) = \prod_{k=0}^{i-1}(x - x_k) \prod_{\ell=0}^{j-1}(y - y_\ell)$$

This basis function is a bivariate polynomial of degree $i + j$.

With $c_{ki,\ell j} = f[x_k, \ldots, x_i][y_\ell, \ldots, y_j]$ we can then write in a purely formal manner [1 pp. 160-164]

$$f(x,y) = \sum_{(i,j)\in\mathbb{N}^2} c_{0i,0j}\; B_{ij}(x,y)$$

The following lemmas about products of basis functions $B_{ij}(x,y)$ and about bivariate divided differences of products of functions will play an important role in the sequel of the text.

Lemma 3.2.

For $k + \ell \geq i + j$ the product $B_{ij}(x,y)\, B_{k\ell}(x,y) =$

$$\sum_{\mu=0}^{i} \sum_{\nu=0}^{j} \lambda_{\mu\nu} B_{k+\mu,\ell+\nu}(x,y)$$

Proof

We write $B_{ij}(x,y) = B_{i0}(x,y)\, B_{0j}(x,y)$.

Since $B_{i0}(x,y)$ is a polynomial in x of degree i we can write

$$B_{i0}(x,y) = \sum_{\mu=0}^{i} \alpha_\mu \left(\prod_{\gamma=k}^{k+\mu-1} (x - x_\gamma) \right)$$

and

$$B_{0j}(x,y) = \sum_{\nu=0}^{j} \beta_\nu \left(\prod_{\gamma=\ell}^{\ell+\nu-1} (y - y_\gamma) \right)$$

with the convention that an empty product is equal to 1. Consequently

$$\begin{aligned} B_{ij}(x,y)\, B_{k\ell}(x,y) &= [B_{k\ell}(x,y)B_{i0}(x,y)]B_{0j}(x,y) \\ &= \left[\sum_{\mu=0}^{i} \alpha_\mu B_{k+\mu,\ell}(x,y) \right] B_{0j}(x,y) \\ &= \sum_{\nu=0}^{j} \sum_{\mu=0}^{i} \alpha_\mu \beta_\nu B_{k+\mu,\ell+\nu}(x,y) \end{aligned}$$

which gives the desired formula if we put $\lambda_{\mu\nu} = \alpha_\mu \beta_\nu$. ∎

A figure in $\mathbb{N}^2$ will clarify the meaning of this lemma. If we multiply $B_{ij}(x, y)$ by $B_{k\ell}(x, y)$ and $k + \ell \geq i + j$ then the only occuring $B_{\mu\nu}(x, y)$ in the product are those with (μ, ν) lying in the shaded rectangle.

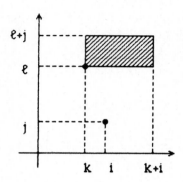

Figure 3.3.

Lemma 3.3.

$$(f \, q)[x_0, \ldots, x_i][y_0, \ldots, y_j]$$
$$= \sum_{\mu=0}^{i} \sum_{\nu=0}^{j} f[x_0, \ldots, x_\mu][y_0, \ldots, y_\nu] \, q[x_\mu, \ldots, x_i][y_\nu, \ldots, y_j]$$

The proof is by induction and analogous to the proof of the univariate case. The definition of multivariate Newton-Padé approximants which we shall give is a very general one. It includes the univariate definition and the multivariate Padé approximants from the previous chapter as a special case as we shall see at the end of this section.

With any finite subset D of $\mathbb{N}^2$ we associate a polynomial

$$\sum_{(i,j)\in D} b_{ij} \, B_{ij} \, (x, y)$$

Given the double Newton series

$$f(x, y) = \sum_{(i,j)\in\mathbb{N}^2} c_{0i,0j} \, B_{ij} \, (x, y)$$

with $c_{0i,0j} = f[x_0, \ldots, x_i][y_0, \ldots, y_j]$, we choose three subsets N, D and E of $\mathbb{N}^2$ and construct an $[N/D]_E$ Newton-Padé approximant to $f(x, y)$ as follows:

$$p(x, y) = \sum_{(i,j) \in N} a_{ij} \, B_{ij} \, (x, y) \quad (N \text{ from ``numerator''}) \qquad (3.22a.)$$

$$q(x, y) = \sum_{(i,j) \in D} b_{ij} \, B_{ij} \, (x, y) \quad (D \text{ from ``denominator''}) \qquad (3.22b.)$$

$$(f \, q - p)(x, y) = \sum_{(i,j) \in \mathbb{N}^2 \backslash E} d_{ij} \, B_{ij} \, (x, y) \quad (E \text{ from ``equations''}) \qquad (3.22c.)$$

We select N, D and E such that

D has $n + 1$ elements, numbered $(i_0, j_0), \ldots, (i_n, j_n)$
$N \subset E$
E satisfies the rectangle rule, i.e. if $(i, j) \in E$ then $(k, \ell) \in E$ for
$$k \leq i \text{ and } \ell \leq j$$
$E \backslash N$ has at least n elements.

Clearly the coefficients d_{ij} in

$$(f \, q - p)(x, y) = \sum_{(i,j) \in \mathbb{N}^2} d_{ij} \, B_{ij}(x, y)$$

are

$$d_{ij} = (f \, q - p)[x_0, \ldots, x_i][y_0, \ldots, y_j]$$

So the conditions (3.22c.) are equivalent with

$$(f \, q - p)[x_0, \ldots, x_i][y_0, \ldots, y_j] = 0 \text{ for } (i, j) \text{ in } E \qquad (3.23.)$$

The system of equations (3.23.) can be divided into a nonhomogeneous and a homogeneous part:

$$(f \, q)[x_0, \ldots, x_i][y_0, \ldots, y_j] = p[x_0, \ldots, x_i][y_0, \ldots, y_j] \text{ for } (i, j) \text{ in } N \qquad (3.23a.)$$
$$(f \, q)[x_0, \ldots, x_i][y_0, \ldots, y_j] = 0 \text{ for } (i, j) \text{ in } E \backslash N \qquad (3.23b.)$$

Let's take a look at the conditions (3.23b.).

Suppose that E is such that exactly n of the homogeneous equations (3.23b.) are linearly independent. We number the respective n elements in $E \backslash N$ with $(h_1, k_1), \ldots, (h_n, k_n)$ and define the set

$$H = \{(h_1, k_1), \ldots, (h_n, k_n)\} \subseteq E \backslash N \qquad \text{(H from "homogeneous equations")}$$

By means of lemma 3.3. we have

$$(f \, q)[x_0, \ldots, x_i][y_0, \ldots, y_j] = (q \, f)[x_0, \ldots, x_i][y_0, \ldots, y_j]$$

$$= \sum_{\mu=0}^{i} \sum_{\nu=0}^{j} q[x_0, \ldots, x_\mu][y_0, \ldots, y_\nu] \, f[x_\mu, \ldots, x_i][y_\nu, \ldots, y_j]$$

Since the only nontrivial $q[x_0, \ldots, x_\mu][y_0, \ldots, y_\nu]$ are the ones with (μ, ν) in D we can write

$$(f \, q)[x_0, \ldots, x_i][y_0, \ldots, y_j] = \sum_{(\mu,\nu) \in D} b_{\mu\nu} \, f[x_\mu, \ldots, x_i][y_\nu, \ldots, y_j]$$

Remember that $f[x_\mu, \ldots, x_i][y_\nu, \ldots, y_j] = 0$ if $\mu > i$ or $\nu > j$. So the homogeneous system of n equations in $n + 1$ unknowns looks like

$$\begin{pmatrix} c_{i_0 h_1, j_0 k_1} & \cdots & c_{i_n h_1, j_n k_1} \\ \vdots & & \vdots \\ c_{i_0 h_n, j_0 k_n} & \cdots & c_{i_n h_n, j_n k_n} \end{pmatrix} \begin{pmatrix} b_{i_0, j_0} \\ \vdots \\ b_{i_n, j_n} \end{pmatrix} = \begin{pmatrix} 0 \\ \vdots \\ 0 \end{pmatrix} \qquad (3.24.)$$

because

$$D = \{(i_0, j_0), \ldots, (i_n, j_n)\}$$

As we suppose the rank of the coefficient matrix to be maximal, a solution $q(x, y)$ is given by

$$q(x,y) = \begin{vmatrix} B_{i_0 j_0}(x,y) & \dots & B_{i_n j_n}(x,y) \\ c_{i_0 h_1, j_0 k_1} & \dots & c_{i_n h_1, j_n k_1} \\ \vdots & & \vdots \\ c_{i_0 h_n, j_0 k_n} & \dots & c_{i_n h_n, j_n k_n} \end{vmatrix} \qquad (3.25a.)$$

By the conditions (3.23a.) and lemma 3.3. we find

$$\begin{aligned} p(x,y) &= \sum_{(i,j)\in N} a_{ij}\ B_{ij}(x,y) \\ &= \sum_{(i,j)\in N} p[x_0,\dots,x_i][y_0,\dots,y_j]\ B_{ij}(x,y) \\ &= \sum_{(i,j)\in N} (q\,f)[x_0,\dots,x_i][y_0,\dots,y_j]\ B_{ij}(x,y) \\ &= \sum_{(\mu,\nu)\in D} b_{\mu\nu}\ \Big(\sum_{(i,j)\in N} c_{\mu i,\nu j}\ B_{ij}(x,y)\Big) \end{aligned}$$

Consequently a determinant representation for $p(x,y)$ is given by

$$p(x,y) = \begin{vmatrix} \displaystyle\sum_{(i,j)\in N} c_{i_0 i, j_0 j} B_{ij}(x,y) & \dots & \displaystyle\sum_{(i,j)\in N} c_{i_n i, j_n j} B_{ij}(x,y) \\ c_{i_0 h_1, j_0 k_1} & \ddots & c_{i_n h_1, j_n k_1} \\ \vdots & & \vdots \\ c_{i_0 h_n, j_0 k_n} & \dots & c_{i_n h_n, j_n k_n} \end{vmatrix} \qquad (3.25b.)$$

If for all $k,\ \ell \geq 0$ we have $q(x_k, y_\ell) \neq 0$ then $\frac{1}{q}(x,y)$ can be written as

$$\frac{1}{q}(x,y) = \sum_{(i,j)\in \mathbb{N}^2} e_{ij}\ B_{ij}\ (x,y)$$

with $e_{ij} = \frac{1}{q}[x_0,\dots,x_i][y_0,\dots,y_j]$. Hence by the use of lemma 3.2. and since E satisfies the inclusion property

$$\Big(f - \frac{p}{q}\Big)(x,y) = \Big[\frac{1}{q}(f\,q - p)\Big](x,y) = \sum_{(i,j)\in \mathbb{N}^2 \backslash E} \tilde{d}_{ij}\ B_{ij}(x,y)$$

The following theorem describes which interpolation conditions are now satisfied by p/q.

Theorem 3.22.

If $q(x_k, y_\ell) \neq 0$ for (k, ℓ) in E then

$$\frac{\partial^{\mu+\nu} f}{\partial x^\mu \partial y^\nu} \ (x_k, y_\ell) = \frac{\partial^{\mu+\nu}(\frac{p}{q})}{\partial x^\mu \partial y^\nu} \ (x_k, y_\ell)$$

for

$$(\mu, \nu) \in I = \{(\mu, \nu) \mid 0 \leq \mu \leq r_k - 1, 0 \leq \nu \leq s_\ell - 1\} \cap \{(\mu, \nu) \mid (k+\mu, \ell+\nu) \in E\}$$

where

$$x_k = x_{k+1} = \ldots = x_{k+r_k-1}$$
$$y_\ell = y_{\ell+1} = \ldots = y_{\ell+s_\ell-1}$$

If $r_k = 1 = s_\ell$ this reduces to

$$f(x_k, y_\ell) = (\frac{p}{q})(x_k, y_\ell) \ \text{ with } \ (k, \ell) \ \text{ in } \ E$$

Proof

Given r_k and s_ℓ for fixed (x_k, y_ℓ), consider the following situation for the interpolation points, with respect to E

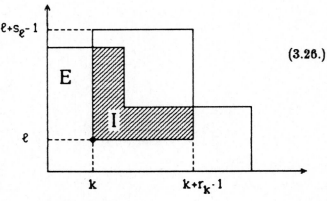

(3.26.)

Figure 3.4.

and define

$$\mu_E = \max\{\mu \mid x_{k+\mu} = x_k \text{ and } (k+\mu, \ell) \in E\}$$
$$\nu_E = \max\{\nu \mid y_{\ell+\nu} = y_\ell \text{ and } (k, \ell+\nu) \in E\}$$
$$\mu_C = \max\{\mu \mid \forall \nu, 0 \le \nu \le \nu_E : (k+\mu, \ell+\nu) \in E\}$$
$$\nu_C = \max\{\nu \mid \forall \mu, 0 \le \mu \le \mu_E : (k+\mu, \ell+\nu) \in E\}$$

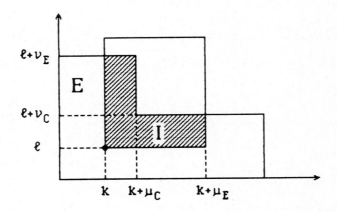

Figure 3.5.

Using these definitions we rewrite I as

$$I = I_1 \cup I_2$$

with

$$I_1 = \{(\mu, \nu) \mid 0 \le \mu \le \mu_E, \ 0 \le \nu \le \nu_C\}$$
$$I_2 = \{(\mu, \nu) \mid 0 \le \mu \le \mu_C, \ 0 \le \nu \le \nu_E\}$$

Because $q(x_k, y_\ell) \neq 0$ for (k, ℓ) in E we have

$$(f - \frac{p}{q})(x,y) = \sum_{(i,j) \in \mathbb{N}^2 \setminus E} \tilde{d}_{ij} \ B_{ij} \ (x,y)$$

To check the interpolation conditions we write

$$\frac{\partial^{\mu+\nu} B_{ij}}{\partial x^{\mu} \partial y^{\nu}} = \frac{\partial^{\mu+\nu} (B_{i0} B_{0j})}{\partial x^{\mu} \partial y^{\nu}}$$

$$= \frac{\partial^{\mu}}{\partial x^{\mu}} \left(\frac{\partial^{\nu}}{\partial y^{\nu}} B_{i0} B_{0j} \right)$$

$$= \frac{\partial^{\mu} B_{i0}}{\partial x^{\mu}} \frac{\partial^{\nu} B_{0j}}{\partial y^{\nu}}$$

If we cover $\mathbb{N}^2 \setminus E$ with three regions

$$A = \{(i,j) \in \mathbb{N}^2 \setminus E \mid i > \mu_E\}$$
$$B = \{(i,j) \in \mathbb{N}^2 \setminus E \mid j > \nu_E\}$$
$$C = \{(i,j) \in \mathbb{N}^2 \setminus E \mid \mu_C < i \le \mu_E, \ \nu_C < j \le \nu_E\}$$

then

$$\left. \frac{\partial^{\mu} B_{i0}}{\partial x^{\mu}} \right|_{(x_k, y_\ell)} = 0 \ \text{ for } (i,j) \text{ in } A \text{ and } (\mu, \nu) \text{ in } I$$

because $B_{i0}(x, y)$ contains a factor $(x - x_k)^{\mu_E + 1}$, and

$$\left. \frac{\partial^{\nu} B_{0j}}{\partial y^{\nu}} \right|_{(x_k, y_\ell)} = 0 \ \text{ for } (i,j) \text{ in } B \text{ and } (\mu, \nu) \text{ in } I$$

because $B_{0j}(x, y)$ contains a factor $(y - y_\ell)^{\nu_E + 1}$.
Analogously

$$\left. \frac{\partial^{\mu} B_{i0}}{\partial x^{\mu}} \right|_{(x_k, y_\ell)} = 0 \ \text{ for } (i,j) \text{ in } C \text{ and } (\mu, \nu) \text{ in } I_2$$

$$\left. \frac{\partial^{\nu} B_{0j}}{\partial y^{\nu}} \right|_{(x_k, y_\ell)} = 0 \ \text{ for } (i,j) \text{ in } C \text{ and } (\mu, \nu) \text{ in } I_1$$

Finally

$$\left. \frac{\partial^{\mu+\nu} (f - \frac{p}{q})}{\partial x^{\mu} \partial y^{\nu}} \right|_{(x_k, y_\ell)} = 0 \ \text{ for } (\mu, \nu) \text{ in } I \text{ and } (k, \ell) \text{ in } E$$

The most general situation for the interpolation points with respect to E is slightly more complicated but completely analogous to the one given in (3.26.). We illustrate this remark by means of the following figure:

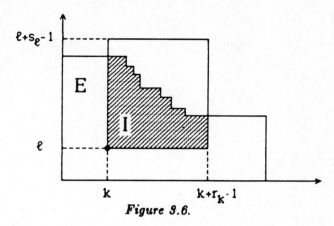

Figure 3.6.

The proof in this case is performed in the same way as above. ∎

We will now obtain the determinant representation given in theorem 3.17. for univariate Newton-Padé approximants, from the determinant representations (3.25a.) and (3.25b.). Consider the Newton interpolating series for $f(x, 0)$ and choose

$$D = \{(j, 0) \mid 0 \leq j \leq n\}$$
$$N = \{(i, 0) \mid 0 \leq i \leq m\}$$
$$E = \{(k, 0) \mid 0 \leq k \leq m + n\}$$

If the points $\{(k, 0) \mid m + 1 \leq k \leq m + n\}$ supply linearly independent equations, then the determinant representations for $p(x, 0)$ and $q(x, 0)$ are

$$q(x, 0) = \begin{vmatrix} 1 & (x - x_0) & \cdots & \prod_{k=0}^{n-1}(x - x_k) \\ c_{0\ m+1,0\ 0} & c_{1\ m+1,0\ 0} & \cdots & c_{n\ m+1,0\ 0} \\ \vdots & \vdots & & \vdots \\ c_{0\ m+n,0\ 0} & c_{1\ m+n,0\ 0} & \cdots & c_{n\ m+n,0\ 0} \end{vmatrix}$$

$$p(x,0) = \begin{vmatrix} \sum_{i=0}^{m} c_{0i,00} \prod_{k=0}^{i-1} (x - x_k) & \cdots & \sum_{i=0}^{m} c_{ni,00} \prod_{k=0}^{i-1} (x - x_k) \\ c_{0\ m+1,0\ 0} & \cdots & c_{n\ m+1,0\ 0} \\ \vdots & & \vdots \\ c_{0\ m+n,0\ 0} & \cdots & c_{n\ m+n,0\ 0} \end{vmatrix}$$

We can also obtain the multivariate Padé approximants defined in the previous chapter as a special case. One only has to choose

$$D = \{(i,j) \mid mn \le i + j \le mn + n\}$$
$$N = \{(i,j) \mid mn \le i + j \le mn + m\}$$
$$E = \{(i,j) \mid mn \le i + j \le mn + m + n\}$$

because when all the interpolation points coincide with the origin, then

$$B_{ij}(x,y) = x^i\,y^j$$

and

$$c_{ih,jk} = \frac{1}{i!}\frac{1}{j!}\frac{\partial^{h-i+k-j} f}{\partial x^{h-i}\ \partial y^{k-j}}\Big|_{(0,0)}$$

Let us now illustrate the multivariate setting by calculating a Newton-Padé approximant for

$$f(x,y) = 1 + \frac{x}{0.1 - y} + \sin(xy)$$

with

$$x_i = i\sqrt{\pi} \qquad\qquad i = 0, 1, 2, \ldots$$
$$y_j = (j - 1)\sqrt{\pi} \qquad j = 0, 1, 2, \ldots$$

The Newton interpolating series looks like

$$f(x,y) = 1 + \frac{1}{0.1 + \sqrt{\pi}} x + \frac{10}{0.1 + \sqrt{\pi}} x(y + \sqrt{\pi})$$

$$+ \frac{10}{0.01 - \pi} x(y + \sqrt{\pi})y + \sum_{i+j \ge 4} c_{0i,0j}\, B_{ij}(x,y)$$

Choose

$$D = \{(0,0), (1,0), (0,1)\}$$
$$N = \{(0,0), (1,0), (0,1), (1,1)\}$$
$$E \backslash N = \{(2,0), (2,1), (0,2), (1,2)\}$$

Writing down the system of equations (3.23b.), it is easy to check that

$$H = \{(2,1), (1,2)\}$$

The determinant formulas for $p(x,y)$ and $q(x,y)$ yield

$$q(x,y) = \begin{vmatrix} 1 & x & y + \sqrt{\pi} \\ c_{02,01} & c_{12,11} & c_{02,11} \\ c_{01,02} & c_{11,02} & c_{01,12} \end{vmatrix}$$

$$= \frac{100}{0.01 - \pi} \left(1 - \frac{1}{0.1 + \sqrt{\pi}}(y + \sqrt{\pi}) \right)$$

$$p(x,y) = \begin{vmatrix} N_{00}(x,y) & N_{10}(x,y) & N_{01}(x,y) \\ c_{02,01} & c_{12,11} & c_{02,11} \\ c_{01,02} & c_{11,02} & c_{01,12} \end{vmatrix}$$

with

$$N_{00}(x,y) = \sum_{i=0}^{1} \sum_{j=0}^{1} c_{0i,0j} B_{ij}(x,y) = 1 + \frac{x}{0.1 + \sqrt{\pi}} + \frac{10}{0.1 + \sqrt{\pi}} x(y + \sqrt{\pi})$$

$$N_{10}(x,y) = \sum_{i=0}^{1} \sum_{j=0}^{1} c_{1i,0j} B_{ij}(x,y) = \frac{0.1 + 2\sqrt{\pi}}{0.1 + \sqrt{\pi}} x + \frac{\sqrt{\pi}}{0.1(0.1 + \sqrt{\pi})} x(y + \sqrt{\pi})$$

$$N_{01}(x,y) = \sum_{i=0}^{1} \sum_{j=0}^{1} c_{0i,1j} B_{ij}(x,y) = (y + \sqrt{\pi}) + 10x(y + \sqrt{\pi})$$

Finally we obtain

$$[N/D]_E(x,y) = \frac{p}{q}(x,y) = \frac{0.1 + \sqrt{\pi} + x - (y + \sqrt{\pi})}{0.1 + \sqrt{\pi} - (y + \sqrt{\pi})}$$

$$= \frac{0.1 + x - y}{0.1 - y}$$

Problems.

(1)   Let

$$r_{m,n} = \frac{p_0}{q_0}$$

be the rational interpolant of order (m,n) for $f(x)$ with $m' = \partial p_0$ and $n' = \partial q_0$. Prove that there exist at least $m' + n' + 1$ points $\{y_1, \ldots, y_s\}$ from the points $\{x_0, \ldots, x_{m+n}\}$ such that $r_{m,n}(y_i) = f(y_i)$ for $i = 1, \ldots, s$.

(2)   Formulate and prove the reciprocal and homografic covariance of rational interpolants.

(3)   Prove theorem 3.8.

(4)   Which interpolation conditions are satisfied by the nth convergent of the continued fraction (3.4) if
a) $d_n = 0$
b) $d_n = \infty$

(5)   Prove theorem 3.17.

(6)   Prove the following result for the error function $(f - r_{m,n})(x)$. Let I be an interval containing all the interpolation points $x_0, \ldots, x_{m+n}$. Then

$$\forall x \in I, \quad \exists y_x \in I :$$

$$(f - r_{m,n})(x) = \frac{1}{(m+n+1)!} \prod_{i=0}^{m+n} (x - x_i)(f - r_{m,n})^{(m+n+1)}(y_x)$$

(7)   Compute $r_{2,2}(x)$ satisfying $r_{2,2}(x_i) = f(x_i)$ for $i = 0, \ldots, 4$ and $r_{1,3}(x)$ satisfying $r_{1,3}(x_i) = f(x_i)$ for $i = 0, \ldots, 4$ with $x_i = i$ $(i = 0, \ldots, 4)$ and $f(x_0) = 4$, $f(x_1) = 2$, $f(x_2) = 1$, $f(x_3) = -1$, $f(x_4) = -4$.

(8)   If $f(x)$ is a rational function

$$f(x) = \frac{r(x)}{s(x)}$$

with $\partial r \leq m$ and $\partial s \leq n$ then $\varphi_k[x_0, \ldots, x_{k-1}, x]$ is constant for a certain k with $0 \leq k \leq 2\max(m,n)$.

(9)   If for some k, $\varphi_k[x_0,\ldots,x_{k-1},x]$ is constant then $f(x)$ is a rational function.

(10) Compute $r_{2,1}(x)$ satisfying $r_{2,1}(x_i) = f(x_i)$ for $i = 0,\ldots,3$ with $x_i = i$ $(i = 0,\ldots,3)$ and $f(x_0) = 1$, $f(x_1) = 3$, $f(x_2) = 2$, $f(x_3) = 4$, by means of
a) the generalized qd-algorithm
b) the algorithm of Gragg

(11) Construct a continued fraction expansion using Thiele's method for $f(x) = \ln(1+x)$ around $x = 0$.

(12) Calculate the inverse differences $\xi_{j,k}^{(1)}$ for $g_1(x,y)$ and $\pi_{j,k}^{(1)}$ for $h_1(x,y)$ and perform one more step in the proof of theorem 3.20. in order to obtain the contribution
$$\frac{(x - x_1)(y - y_1)}{B_2(x,y)}$$

with

$$B_2(x,y) = \varphi_2[x_0,x_1,x_2][y_0,y_1,y_2]$$
$$+ \sum_{k=3}^{\infty} \left| \frac{x - x_{k-1}}{\varphi_{k,2}[x_0,\ldots,x_k][y_0,y_1,y_2]} \right|$$
$$+ \sum_{k=3}^{\infty} \left| \frac{y - y_{k-1}}{\varphi_{2,k}[x_0,x_1,x_2][y_0,\ldots,y_k]} \right|$$

in the continued fraction (3.20.).

Remarks.

(1)   Instead of polynomials

$$p(x) = \sum_{i=0}^{m} a_i \, x^i$$

and

$$q(x) = \sum_{i=0}^{n} b_i \, x^i$$

one could also use linear combinations

$$p(x) = \sum_{i=0}^{m} a_i \, g_i(x)$$

and

$$q(x) = \sum_{i=0}^{n} b_i \, g_i(x)$$

of basis functions $\{g_i\}_{i \in \mathbb{N}}$, which we call **generalized polynomials**, and study the generalized rational interpolation problem

$$(f \, q - p)(x_i) = 0 \qquad i = 0, \ldots, m + n$$

A unique solution of this interpolation problem exists provided $\{g_i(x)\}_{i \in \mathbb{N}}$ satisfies the Haar condition, i.e. for every $k \geq 0$ and for every set of distinct points $\{x_0, \ldots, x_k\}$ the generalized Vandermonde determinant

$$\begin{vmatrix} g_0(x_0) & \cdots & g_k(x_0) \\ \vdots & & \vdots \\ g_0(x_k) & \cdots & g_k(x_k) \end{vmatrix} \neq 0$$

Examples of such interpolation problems can be found in [15]. A recursive algorithm for the calculation of these **generalized rational interpolants** is given in [13].

(2) The Newton-Padé approximation problem (3.15.) is a linear problem in that sense that $r_{m,n}$ can be considered as the root of the linear equation

$$q\, r_{m,n} - p = 0$$

where p and q are determined by the following interpolation conditions

$$(q\, f - p)(x_j) = 0 \qquad j = 0, \ldots, m + n$$

Instead of such linear equations one can also consider algebraic equations

$$\sum_{i=0}^{k} r^i_{m_0, \ldots, m_k} p_i = 0$$

where the polynomials p_i of degree m_i are determined by

$$\sum_{i=0}^{k} f^i(x_j) p_i(x_j) = 0 \qquad j = 0, \ldots, m_0 + \ldots + m_k + k - 1$$

More generally we can consider for different functions $f_0(x), \ldots, f_k(x)$ the interpolation conditions

$$\sum_{i=0}^{k} f_i(x_j) p_i(x_j) = 0 \qquad j = 0, \ldots, m_0 + \ldots + m_k + k - 1$$

An extensive study of this type of problems is made in [14] and [10].

(3) Rational interpolants have also been defined for vector valued functions [11, 25] using generalized vector inverses. For other definitions of multivariate rational interpolants we refer to [17] and [8]: Siemaszko uses nonsymmetric branched continued fractions while in [8] Stoer's recursive scheme for the calculation of univariate rational interpolants is generalized to the multivariate case.

References.

[1] *Berezin J.* and *Zhidkov N.*   Computing methods I. Addison Wesley, New York, 1965.

[2] *Claessens G.*   A generalization of the qd-algorithm. J. Comput. Appl. Math. 7, 1981, 237-247.

[3] *Claessens G.*   A new algorithm for osculatory rational interpolation. Numer. Math. 27, 1976, 77-83.

[4] *Claessens G.*   Some aspects of the rational Hermite interpolation table and its applications. Ph. D., University of Antwerp, 1976.

[5] *Claessens G.*   A useful identity for the rational Hermite interpolation table. Numer. Math. 29, 1978, 227-231.

[6] *Claessens G.*   On the Newton-Padé approximation problem. J. Approx. Theory 22, 1978, 150-160.

[7] *Cuyt A.* and *Verdonk B.*   General order Newton-Padé approximants for multivariate functions. Numer. Math. 43, 1984, 293-307.

[8] *Cuyt A.* and *Verdonk B.*   Multivariate rational interpolation. Computing 34, 1985, 41-61.

[9] *Davis Ph.*   Interpolation and approximation. Blaisdell, New York, 1965.

[10] *Della Dora J.*   Contribution à l'approximation de fonctions de la variable complexe au sens Hermite-Padé et de Hardy. Ph. D., University of Grenoble, 1980.

[11] *Graves-Morris P.* and *Jenkins C.*   Generalised inverse vector valued rational interpolation. In [22], 144-156.

[12] *Larkin F.*   Some techniques for rational interpolation. Comput. J. 10, 1967, 178-187.

[13] *Loi S.* and *Mc Innes A.*   An algorithm for generalized rational interpolation. BIT 23, 1983, 105-117.

[14] *Lübbe W.*   Ueber ein allgemeines Interpolationsproblem und lineare Identitäten zwischen benachbarten Lösungssystemen. Ph. D., University of Hannover, 1983.

[15] *Mühlbach G.*   The general Neville-Aitken algorithm and some applications. Numer. Math. 31, 1978, 97-110.

[16] *Salzer H.*   Note on osculatory rational interpolation. Math. Comp. 16, 1962, 486-491.

[17] *Siemaszko W.*   Thiele-type branched continued fractions for two-variable functions. J. Comput. Appl. Math. 9, 1983, 137-153.

[18] *Stoer J.*   Ueber zwei Algorithmen zur Interpolation mit rationalen Funktionen. Numer. Math. 3, 1961, 285-304.

[19] *Thiele T.*   Interpolationsrechnung. Teubner, Leipzig, 1909.

[20] *Walsh J.*   Interpolation and approximation by rational functions in the complex domain. Amer. Math. Soc., Providence Rhode Island, 1969.

[21] *Warner D.*   Hermite interpolation with rational functions. Ph. D., University of California, 1974.

[22] *Werner H.* and *Bünger H.*   Padé approximation and its applications. Lecture Notes in Mathematics 1071, Springer, Berlin, 1984.

[23] *Wuytack L.*   On some aspects of the rational interpolation problem. SIAM. J. Numer. Anal. 11, 1974, 52-60.

[24] *Wuytack L.*   On the osculatory rational interpolation problem. Math. Comp. 29, 1975, 837 - 843.

[25] *Wynn P.*   Continued fractions whose coefficients obey a non-commutative law of multplication. Arch. Rational Mech. Anal. 12, 1963, 273-312.

[26] *Wynn P.*   Ueber einen Interpolations-Algorithmus und gewisse andere Formeln, die in der Theorie der Interpolation durch rationale Funktionen bestehen. Numer. Math. 2, 1960, 151-182.

CHAPTER IV: Applications.

"It is my hope that by demonstrating the ease with which the various transformations may be effected, their field of application might be widened, and deeper insight thereby obtained into the problems for whose solution the transformations have been used."

P. WYNN — *"On a device for computing the $e_m(S_n)$ transformation"* *(1956).*

The approximations introduced in the previous chapters will now be used to develop techniques for the solution of various mathematical problems: convergence acceleration, numerical integration, the solution of one or more simultaneous nonlinear equations, the solution of initial value problems, boundary value problems, partial differential equations, integral equations, etc. Since these techniques are based on nonlinear approximations they shall be nonlinear themselves. We shall discuss advantages and disadvantages in each of the sections separately and illustrate their use by means of numerical examples.

§1. Convergence acceleration.

1.1. The univariate ε-algorithm.

Consider a sequence $\{a_i\}_{i \in \mathbb{N}}$ of real or complex numbers with

$$\lim_{i \to \infty} a_i = A$$

Since we are interested in the limiting value A of the sequence we shall try to construct a sequence $\{b_i\}_{i \in \mathbb{N}}$ that converges faster to A, or

$$\lim_{i \to \infty} \frac{|b_i - A|}{|a_i - A|} = 0$$

We shall describe here some nonlinear techniques that can be used for the construction of $\{b_i\}_{i \in \mathbb{N}}$.
Consider the univariate power series

$$f(x) = a_0 + \sum_{i=1}^{\infty} \nabla a_i \ x^i$$

with $\nabla a_i = a_i - a_{i-1}$. Then clearly for the partial sums

$$F_k(x) = a_0 + \sum_{i=1}^{k} \nabla a_i \ x^i$$

we have

$$F_k(1) = a_k \qquad\qquad k = 0, 1, 2, \ldots$$

If we approximate $f(x)$ by $r_{i,i}(x)$, the Padé approximant of order (i,i) for f, then we can put

$$b_i = r_{i,i}(1) \qquad\qquad i = 0, 1, 2, \ldots$$

For the computation of b_i the ϵ-algorithm can be used:

$$\epsilon_{-1}^{(k)} = 0 \qquad\qquad\qquad\qquad k = 0, 1, \ldots$$

$$\epsilon_{0}^{(k)} = F_k(1) = a_k \qquad\qquad\quad k = 0, 1, \ldots$$

$$\epsilon_{k+1}^{(\ell)} = \epsilon_{k-1}^{(\ell+1)} + \frac{1}{\epsilon_{k}^{(\ell+1)} - \epsilon_{k}^{(\ell)}} \qquad k = 0, 1, \ldots \qquad \ell = 0, 1, \ldots$$

Then

$$\epsilon_{2i}^{(0)} = r_{i,i}(1) = b_i$$

As described in section 3 of chapter II these ϵ-values can be ordered in a table

$$
\begin{array}{ccccccc}
\epsilon_{-1}^{(0)} \\
 & \epsilon_{0}^{(0)} \\
\epsilon_{-1}^{(1)} & & \epsilon_{1}^{(0)} \\
 & \epsilon_{0}^{(1)} & & \epsilon_{2}^{(0)} \\
\epsilon_{-1}^{(2)} & & \epsilon_{1}^{(1)} & & \epsilon_{3}^{(0)} \\
 & \epsilon_{0}^{(2)} & & \epsilon_{2}^{(1)} & & \ddots \\
\epsilon_{-1}^{(3)} & \vdots & \epsilon_{1}^{(2)} & \vdots & \epsilon_{3}^{(1)} & & \cdots \\
\vdots & & \vdots & & \vdots
\end{array}
$$

The convergence of the sequence $\{b_i\}_{i \in \mathbb{N}}$ depends very much on the given sequence $\{a_i\}_{i \in \mathbb{N}}$. Of course the convergence properties of $\{b_i\}_{i \in \mathbb{N}}$ are the same as those of the diagonal Padé approximants evaluated at $x = 1$ and for this we refer to section 4 of chapter II. In some special cases it is possible to prove acceleration of the convergence of $\{a_i\}_{i \in \mathbb{N}}$.

A sequence $\{a_i\}_{i \in \mathbb{N}}$ is called **totally monotone** if

$$(-1)^k \; \Delta^k \; a_i \geq 0 \qquad i, k = 0, 1, 2, \ldots$$

where $\Delta^k a_i = \Delta^{k-1} a_{i+1} - \Delta^{k-1} a_i$ and $\Delta^0 a_i = a_i$. In other words, $\{a_i\}_{i \in \mathbb{N}}$ is totally monotone if

$$a_0 \geq a_1 \geq a_2 \geq \ldots \geq 0$$
$$\Delta a_0 \leq \Delta a_1 \leq \Delta a_2 \leq \ldots \leq 0$$
$$\Delta^2 a_0 \geq \Delta^2 a_1 \geq \Delta^2 a_2 \geq \ldots \geq 0$$

and so on.

Note that every totally monotone sequence actually converges.
The sequences $\{\lambda^i\}_{i\in\mathbb{N}}$ for $0 \leq \lambda \leq 1$ and $\{1/(i+1)\}_{i\in\mathbb{N}}$ are for instance totally monotone sequences. The close link with the theory of Stieltjes series becomes clear in the following theorem [4 p.81].

Theorem 4.1.

The sequence $\{a_i\}_{i\in\mathbb{N}}$ is totally monotone if and only if there exists a real-valued, bounded and nondecreasing function g on $[0,1]$ taking on infinitely many different values such that

$$a_i = \int\limits_0^1 t^i \; dg(t) \quad i = 0, 1, 2, \ldots$$

A sequence $\{a_i\}_{i\in\mathbb{N}}$ is called **totally oscillating** if the sequence $\{(-1)^i \; a_i\}_{i\in\mathbb{N}}$ is totally monotone. One can see that every convergent totally oscillating sequence converges to zero.
For these sequences the following results can be proved [4 pp. 83-85].

Theorem 4.2.

If the ϵ-algorithm is applied to the sequence $\{a_i\}_{i\in\mathbb{N}}$ with $\lim_{i\to\infty} a_i = A$ and if there exist constants α and β such that $\{\alpha a_i + \beta\}_{i\in\mathbb{N}}$ is a totally monotone sequence, then

$$\lim_{\ell\to\infty} \; \epsilon_{2k}^{(\ell)} = A \; \; k \geq 0 \text{ and fixed}$$

$$\lim_{k\to\infty} \; \epsilon_{2k}^{(\ell)} = A \; \; \ell \geq 0 \text{ and fixed}$$

If also

$$\lim_{i\to\infty} \; \frac{a_{i+1} - A}{a_i - A} \neq 1$$

then

$$\lim_{\ell\to\infty} \; \frac{\epsilon_{2k}^{(\ell)} - A}{a_{2k+\ell} - A} = 0 \; \; k \geq 0 \text{ and fixed}$$

$$\lim_{k\to\infty} \; \frac{\epsilon_{2k}^{(\ell)} - A}{a_{2k+\ell} - A} = 0 \; \; \ell \geq 0 \text{ and fixed}$$

Theorem 4.3.

If the ϵ-algorithm is applied to the sequence $\{a_i\}_{i\in\mathbb{N}}$ with $\lim_{i\to\infty} a_i = A = 0$ and if there exist constants α and β such that $\{\alpha a_i + \beta\}_{i\in\mathbb{N}}$ is a totally oscillating sequence, then

$$\lim_{\ell\to\infty}\quad \epsilon_{2k}^{(\ell)} = 0 \quad k \geq 0, \text{fixed}$$

$$\lim_{k\to\infty}\quad \epsilon_{2k}^{(\ell)} = 0 \quad \ell \geq 0, \text{fixed}$$

$$\lim_{\ell\to\infty}\quad \frac{\epsilon_{2k}^{(\ell)} - A}{a_{2k+\ell} - A} = 0 \quad k \geq 0, \text{fixed}$$

$$\lim_{k\to\infty}\quad \frac{\epsilon_{2k}^{(\ell)} - A}{a_{2k+\ell} - A} = 0 \quad \ell \geq 0, \text{fixed}$$

We illustrate these theorems with some numerical results. Consider

$$\{a_i\}_{i\in\mathbb{N}} = \{\frac{1}{i+1}\}_{i\in\mathbb{N}}$$

This is a totally monotone sequence. In table 4.1. we have listed the values $\epsilon_{2n}^{(m-n)}$ for $n = 0,\ldots,4$ and $m = 0,\ldots,10$. So the values $b_i = \epsilon_{2i}^{(0)}$ can be found on the main diagonal. Clearly the convergence of the sequence $\{a_i\}_{i\in\mathbb{N}}$ is accelerated. All these computations were performed using double precision arithmetic (56–digit binary representation). To illustrate the influence of data perturbations and rounding errors we give in table 4.2. the same values $\epsilon_{2n}^{(m-n)}$ obtained using single precision input and single precision arithmetic (24–digit binary representation). In the tables 4.3. and 4.4. the respective results are given for the totally oscillating sequence

$$\{a_i\}_{i\in\mathbb{N}} = \{\frac{(-1)^i}{i+1}\}_{i\in\mathbb{N}}$$

Here the convergence of the sequence $\{b_i\}_{i\in\mathbb{N}}$ is much faster than that of the sequence $\{a_i\}_{i\in\mathbb{N}}$.

These examples illustrate that the influence of data perturbations and rounding errors can be important. When $\epsilon_k^{(\ell+1)}$ is nearly equal to $\epsilon_k^{(\ell)}$ one can loose a lot of significant digits. In [62] it is shown how this misfortune may be overcome.

Table 4.1.

$$\epsilon_{2n}^{(m-n)} \ for \ \epsilon_0^{(m)} = a_m = \frac{1}{m+1}$$

56-digit binary representation

m\n	0	1	2	3	4
0	0.10000000D+01	0.66666667D+00	0.52173913D+00	0.43636364D+00	0.37874803D+00
1	0.50000000D+00	0.25000000D+00	0.17647059D+00	0.13801453D+00	0.11398283D+00
2	0.33333333D+00	0.16666667D+00	0.11111111D+00	0.85271318D-01	0.69561659D-01
3	0.25000000D+00	0.12500000D+00	0.83333333D-01	0.62500000D-01	0.50607287D-01
4	0.20000000D+00	0.10000000D+00	0.66666667D-01	0.50000000D-01	0.40000000D-01
5	0.16666667D+00	0.83333333D-01	0.55555556D-01	0.41666667D-01	0.33333333D-01
6	0.14285714D+00	0.71428571D-01	0.47619048D-01	0.35714286D-01	0.28571429D-01
7	0.12500000D+00	0.62500000D-01	0.41666667D-01	0.31250000D-01	0.25000000D-01
8	0.11111111D+00	0.55555556D-01	0.37037037D-01	0.27777778D-01	0.22222222D-01
9	0.10000000D+00	0.50000000D-01	0.33333333D-01	0.25000000D-01	0.20000000D-01
10	0.90909091D-01	0.45454545D-01	0.30303030D-01	0.22727273D-01	0.18181818D-01

Table 4.2.

$$\epsilon_{2n}^{(m-n)} \ for \ \epsilon_0^{(m)} = a_m = \frac{1}{m+1}$$

24-digit binary representation

m\n	0	1	2	3	4
0	0.10000000E+01	0.66666669E+00	0.52173918E+00	0.43636370E+00	0.37874800E+00
1	0.50000000E+00	0.25000008E+00	0.17647058E+00	0.13801455E+00	0.11398285E+00
2	0.33333334E+00	0.16666664E+00	0.11111123E+00	0.85271232E-01	0.69561742E-01
3	0.25000000E+00	0.12500004E+00	0.83333202E-01	0.62500395E-01	0.50605543E-01
4	0.20000000E+00	0.10000000E+00	0.66666812E-01	0.49997989E-01	0.40013745E-01
5	0.16666667E+00	0.83333351E-01	0.55554848E-01	0.41675072E-01	0.33280782E-01
6	0.14285715E+00	0.71428463E-01	0.47620041E-01	0.35695387E-01	0.28692538E-01
7	0.12500000E+00	0.62500104E-01	0.41664694E-01	0.31274047E-01	0.24795113E-01
8	0.11111111E+00	0.55555556E-01	0.37037980E-01	0.27756441E-01	0.22477847E-01
9	0.10000000E+00	0.50000019E-01	0.33332549E-01	0.25024110E-01	0.19564494E-01
10	0.90909094E-01	0.45454517E-01	0.30304417E-01	0.22691974E-01	0.18706985E-01

Table 4.3.

$$\epsilon_{2n}^{(m-n)} \quad for \quad \epsilon_0^{(m)} = a_m = \frac{(-1)^m}{m+1}$$

56-digit binary representation

m\n	0	1	2	3	4
0	0.10000000D+01	0.40000000D+00	0.25531915D+00	0.18004651D+00	0.14690879D+00
1	−0.50000000D+00	0.35714286D−01	−0.21897810D−01	0.11833091D−01	−0.96169448D−02
2	0.33333333D+00	−0.98039216D−02	0.11454754D−02	−0.36107008D−03	0.15341245D−03
3	−0.25000000D+00	0.40322581D−02	−0.25960540D−03	0.34896706D−04	−0.89086733D−05
4	0.20000000D+00	−0.20408163D−02	0.83229297D−04	−0.73003358D−05	0.10459978D−05
5	−0.16666667D+00	0.11737089D−02	−0.33049111D−04	0.20314308D−05	−0.20944207D−06
6	0.14285714D+00	−0.73637703D−03	0.15179805D−04	−0.68891969D−06	0.53279358D−07
7	−0.12500000D+00	0.49212598D−03	−0.77490546D−05	0.27006940D−06	−0.16203117D−07
8	0.11111111D+00	−0.34506556D−03	0.42861980D−05	−0.11826320D−06	0.56560969D−08
9	−0.10000000D+00	0.25125628D−03	−0.25250612D−05	0.56510073D−07	−0.22028952D−08
10	0.90909091D−01	−0.18860807D−03	0.15651583D−05	−0.28977004D−07	0.93783049D−09

Table 4.4.

$$\epsilon_{2n}^{(m-n)} \quad for \quad \epsilon_0^{(m)} = a_m = \frac{(-1)^m}{m+1}$$

24-digit binary representation

$m \backslash n$	0	1	2	3	4
0	0.10000000E+01	0.40000004E+00	0.25531918E+00	0.18604654E+00	0.14690882E+00
1	−0.50000000E+00	0.35714284E−01	−0.21897800E−01	0.11833105E−01	−0.96169570E−02
2	0.33333334E+00	−0.98039182E−02	0.11454860E−02	−0.36106500E−03	0.15340716E−03
3	−0.25000000E+00	0.40322733E−02	−0.25959974E−03	0.34896013E−04	−0.89125078E−05
4	0.20000000E+00	−0.20408179E−02	0.83226092E−04	−0.73038805E−05	0.10442981E−05
5	−0.16666667E+00	0.11736987E−02	−0.33053879E−04	0.20300561E−05	−0.20829127E−06
6	0.14285715E+00	−0.73687674E−03	0.15181173E−04	−0.68641026E−06	0.56101026E−07
7	−0.12500000E+00	0.49213349E−03	−0.77439190E−05	0.27358018E−06	−0.13825897E−07
8	0.11111111E+00	−0.34506133E−03	0.42287141E−05	−0.11667172E−06	0.66095391E−08
9	−0.10000000E+00	0.25125383E−03	−0.25255961E−05	0.56633620E−07	−0.17431754E−08
10	0.90909094E−01	−0.18860842E−03	0.15651756E−05	−0.28493092E−07	0.16364775E−08

Various generalizations to accelerate the convergence of a sequence of vectors, matrices, and so on exist. We refer to [4] and [61]. The convergence of a multi-dimensional table of real or complex numbers is treated in section 1.5.

1.2. The qd-algorithm.

To accelerate the convergence of $\{a_i\}_{i\in\mathbb{N}}$ we construct again

$$f(x) = a_0 + \sum_{i=1}^{\infty} \nabla a_i \, x^i$$

We have seen in section 3 of chapter II that it is possible to construct a corresponding continued fraction

$$d_0 + \frac{d_1 x}{\vert 1} + \frac{d_2 x}{\vert 1} + \frac{d_3 x}{\vert 1} + \ldots$$

where the coefficients d_i can be calculated by means of the qd-algorithm. If we calculate the ith convergent $C_i(x)$ of this continued fraction, we can now put

$$b_i = C_i(1) \qquad i = 0, 1, 2, \ldots$$

The numbers b_i can be computed using a continued fraction algorithm from section 5 of chapter I. Remark that theoretically the numbers b_{2i} here are the same as the b_i computed by means of the ϵ-algorithm. In practice there are however differences due to rounding errors. We will illustrate this numerically. The tables 4.5. and 4.6. illustrate the effect of perturbations when the qd-table is computed: the results in table 4.5. are obtained using double precision arithmetic while those in table 4.6. are obtained using single precision arithmetic. Input was

$$\{a_i\}_{i\in\mathbb{N}} = \{\sum_{j=0}^{i} \frac{1}{j!}\}_{i\in\mathbb{N}}$$

So $f(x) = e^x$ and $\lim_{i\to\infty} a_i = e = 2.718281828\ldots$
From the tables 4.5. and 4.6. the values b_i were computed using the three-term recurrence relations (1.3.) respectively in double and single precision. The results can be found in the respective tables 4.7. and 4.8., where they can also be compared with the values of the ϵ-algorithm.

Table 4.5.

$$qd - \text{table for } q_1^{(k)} = \frac{\nabla a_{k+1}}{\nabla a_k} \text{ with } a_k = \sum_{j=0}^{k} \frac{1}{j!}, \qquad k = 1, 2, \ldots$$

56−digit binary representation

$q_1^{(k)}$	$e_1^{(k)}$	$q_2^{(k)}$	$e_2^{(k)}$	$q_3^{(k)}$	$e_3^{(k)}$	$q_4^{(k)}$	$e_4^{(k)}$	$q_5^{(k)}$	$e_5^{(k)}$
0.50000000D+00									
	−0.16666667D+00								
0.33333333D+00		0.16666667D+00							
	−0.83333333D-01		−0.66666667D-01						
0.25000000D+00		0.15000000D+00		0.10000000D+00					
	−0.50000000D-01		−0.47619048D-01		−0.71428571D-01				
0.20000000D+00		0.13333333D+00		0.95238095D-01		0.71428571D-01			
	−0.33333333D-01		−0.35714286D-01		−0.53571429D-01		−0.55555556D-01		
0.16666667D+00		0.11904762D+00		0.89285714D-01		0.69444444D-01		0.55555556D-01	
	−0.23809524D-01		−0.27777778D-01		−0.41666667D-01		−0.44444444D-01		−0.45454545D-01
0.14285714D+00		0.10714286D+00		0.83333333D-01		0.66666667D-01		0.54545455D-01	
	−0.17857143D-01		−0.22222222D-01		−0.33333333D-01		−0.36363636D-01		
0.12500000D+00		0.97222222D-01		0.77777778D-01		0.63636364D-01			
	−0.13888889D-01		−0.18181818D-01		−0.27272727D-01				
0.11111111D+00		0.88888889D-01		0.72727273D-01					
	−0.11111111D-01								
0.10000000D+00		0.81818182D-01							
	−0.90909091D-02								
0.90909091D-01									

Table 4.6.

$$qd - \text{table for } q_1^{(k)} = \frac{\nabla a_{k+1}}{\nabla a_k} \ \text{ with } a_k = \sum_{j=0}^{k} \frac{1}{j!} \qquad k = 1, 2, \ldots$$

24−digit binary representation

$q_1^{(k)}$	$e_1^{(k)}$	$q_2^{(k)}$	$e_2^{(k)}$	$q_3^{(k)}$	$e_3^{(k)}$	$q_4^{(k)}$	$e_4^{(k)}$	$q_5^{(k)}$	$e_5^{(k)}$
0.50000000E+00									
	−0.16666666E+00								
0.33333334E+00		0.16666670E+00							
	−0.83333343E−01		−0.10000007E+00						
0.25000000E+00		0.14999998E+00		0.99999800E−01					
	−0.49999990E−01		−0.66666588E−01		−0.71427882E−01				
0.20000000E+00		0.13333338E+00		0.95238507E−01		0.71430743E−01			
	−0.33333346E−01		−0.47619179E−01		−0.53572312E−01		−0.55559866E−01		
0.16666666E+00		0.11904755E+00		0.89285374E−01		0.69443189E−01		0.55554055E−01	
	−0.23809522E−01		−0.35714269E−01		−0.41666761E−01		−0.44447407E−01		−0.45489557E−01
0.14285713E+00		0.10714281E+00		0.83332881E−01		0.66662453E−01		0.54511994E−01	
	−0.17857134E−01		−0.27777627E−01		−0.33331484E−01		−0.36346123E−01		
0.12500000E+00		0.07222313E−01		0.77779025E−01		0.63647814E−01			
	−0.13888896E−01		−0.22222437E−01		−0.27275681E−01				
0.11111110E+00		0.88888772E−01		0.72725780E−01					
	−0.11111103E−01		−0.18181644E−01						
0.10000000E+00		0.81818230E−01							
	−0.90000079E−02								
0.00000094E−01									

Table 4.7.

$$a_i = \sum_{j=0}^{i} \frac{1}{j!} \qquad \lim_{i \to \infty} a_i = 2.71828182845904\ldots$$

56-digit binary representation

i	$C_{2i}(1)$	$\epsilon_{2i}^{(0)}$
0	1.000000000000000D+00	1.000000000000000D+00
1	3.000000000000000D+00	3.000000000000000D+00
2	2.714285714285714D+00	2.714285714285714D+00
3	2.718309859154930D+00	2.718309859154930D+00
4	2.718281718281718D+00	2.718281718281718D+00
5	2.718281828735696D+00	2.718281828735696D+00
6	2.718281828458563D+00	2.718281828458564D+00
7	2.718281828459046D+00	2.718281828459046D+00
8	2.718281828459045D+00	2.718281828459045D+00

Table 4.8.

$$a_i = \sum_{j=0}^{i} \frac{1}{j!} \qquad \lim_{i \to \infty} a_i = 2.718281828\ldots$$

24-digit binary representation

i	$C_{2i}(1)$	$\epsilon_{2i}^{(0)}$
0	1.0000000E+00	1.0000000E+00
1	3.0000000E+00	3.0000000E+00
2	2.7142858E+00	2.7142858E+00
3	2.7183099E+00	2.7183099E+00
4	2.7182818E+00	2.7182818E+00
5	2.7182815E+00	2.7182822E+00

1.3. The algorithm of Bulirsch-Stoer.

Let $\{x_i\}_{i \in \mathbb{N}}$ be a convergent sequence of points with

$$\lim_{i \to \infty} x_i = z$$

When using extrapolation techniques to accelerate the convergence of $\{a_i\}_{i \in \mathbb{N}}$ we compute a sequence $\{b_i\}_{i \in \mathbb{N}}$ with

$$b_i = \lim_{x \to z} g_i(x)$$

where $g_i(x)$ is determined by the interpolation conditions

$$g_i(x_j) = a_j \quad j = 0,\dots,i \text{ with } i \geq 0$$

The point z is called the **extrapolation point**. Often polynomials are used for the interpolating functions g_i. We consider here the case of rational interpolation where we shall use Stoer's recursive scheme given in section 3 of chapter III. Using the notations of chapter III and writing

$$r_{m,n}^{(j)} = \frac{p_{m,n}^{(j)}}{q_{m,n}^{(j)}}$$

we first generalize the formulas (3.12.):

$$p_{m,n}^{(j)}(x) = (x - x_j) \, b_{m-1,n}^{(j)} \, p_{m-1,n}^{(j+1)}(x) - (x - x_{j+m+n}) \, b_{m-1,n}^{(j+1)} \, p_{m-1,n}^{(j)}(x)$$
$$q_{m,n}^{(j)}(x) = (x - x_j) \, b_{m-1,n}^{(j)} \, q_{m-1,n}^{(j+1)}(x) - (x - x_{j+m+n}) \, b_{m-1,n}^{(j+1)} \, q_{m-1,n}^{(j)}(x)$$

and

$$p_{m,n}^{(j)}(x) = (x - x_j) \, a_{m,n-1}^{(j)} \, p_{m,n-1}^{(j+1)}(x) - (x - x_{j+m+n}) \, a_{m,n-1}^{(j+1)} \, p_{m,n-1}^{(j)}(x)$$
$$q_{m,n}^{(j)}(x) = (x - x_j) \, a_{m,n-1}^{(j)} \, q_{m,n-1}^{(j+1)}(x) - (x - x_{j+m+n}) \, a_{m,n-1}^{(j+1)} \, q_{m,n-1}^{(j)}(x)$$

with $p_{0,0}^{(j)}(x) = a_j$ and $q_{0,0}^{(j)}(x) = 1$.

We can easily see taking into account the interpolation properties of $r_{m,n}^{(j)}$ and $r_{m,n-1}^{(j+1)}$ that

$$r_{m,n}^{(j)}(x) - r_{m,n-1}^{(j+1)}(x)$$

$$= \left(-a_{m,n-1}^{(j+1)} \, b_{m,n}^{(j)}\right) \frac{\prod_{k=1}^{m+n}(x - x_{j+k})}{q_{m,n}^{(j)}(x) q_{m,n-1}^{(j+1)}(x)} \tag{4.1a.}$$

and analogously that

$$r_{m,n}^{(j+1)}(x) - r_{m,n-1}^{(j+1)}(x)$$
$$= \left(-a_{m,n-1}^{(j+1)}\ b_{m,n}^{(j+1)}\right)\frac{\prod_{k=1}^{m+n}(x - x_{j+k})}{q_{m,n}^{(j+1)}(x)q_{m,n-1}^{(j+1)}(x)} \qquad (4.1b.)$$

If $a_{m,n-1}^{(j+1)} \neq 0$, then we can write

$$r_{m+1,n}^{(j)}(x) = \frac{a_{m,n-1}^{(j+1)}\left[(x - x_j)\ b_{m,n}^{(j)}\ p_{m,n}^{(j+1)}(x) - (x - x_{j+m+n+1})\ b_{m,n}^{(j+1)}\ p_{m,n}^{(j)}(x)\right]}{a_{m,n-1}^{(j+1)}\left[(x - x_j)\ b_{m,n}^{(j)}\ q_{m,n}^{(j+1)}(x) - (x - x_{j+m+n+1})\ b_{m,n}^{(j+1)}\ q_{m,n}^{(j)}(x)\right]}$$

and thus, using (4.1.),

$$r_{m+1,n}^{(j)}(x)$$
$$= \frac{(x - x_j)\ r_{m,n}^{(j+1)}\left[r_{m,n-1}^{(j+1)} - r_{m,n}^{(j)}\right] - (x - x_{j+m+n+1})\ r_{m,n}^{(j)}\left[r_{m,n-1}^{(j+1)} - r_{m,n}^{(j+1)}\right]}{(x - x_j)\left[r_{m,n-1}^{(j+1)} - r_{m,n}^{(j)}\right] - (x - x_{j+m+n+1})\left[r_{m,n-1}^{(j+1)} - r_{m,n}^{(j+1)}\right]}$$

or

$$r_{m+1,n}^{(j)}(x)$$
$$= r_{m,n}^{(j)} + \frac{r_{m,n}^{(j+1)}(x) - r_{m,n}^{(j)}(x)}{1 - \dfrac{x - x_{j+m+n+1}}{x - x_j}\left(\dfrac{r_{m,n}^{(j+1)}(x) - r_{m,n-1}^{(j+1)}(x)}{r_{m,n}^{(j)}(x) - r_{m,n-1}^{(j+1)}(x)}\right)}$$

Analogously

$$r_{m,n+1}^{(j)}(x) = r_{m,n}^{(j)}(x) + \frac{r_{m,n}^{(j+1)}(x) - r_{m,n}^{(j)}(x)}{1 - \dfrac{x - x_{j+m+n+1}}{x - x_j}\left(\dfrac{r_{m,n}^{(j+1)}(x) - r_{m-1,n}^{(j+1)}(x)}{r_{m,n}^{(j)}(x) - r_{m-1,n}^{(j+1)}(x)}\right)}$$

If we use the interpolating functions

$$\{g_i(x)\}_{i \in \mathbb{N}} = \{r_{0,0}^{(0)}(x), r_{0,1}^{(0)}(x), r_{1,1}^{(0)}(x), r_{1,2}^{(0)}(x), \ldots\}$$

and write $s_i^{(j)} = r_{m,n}^{(j)}(z)$ for $m + n = i$, then

$$s_{i+1}^{(j)} = s_i^{(j)} + \frac{s_i^{(j+1)} - s_i^{(j)}}{1 - \left(\dfrac{x_{j+i+1} - z}{x_j - z}\right)\left(1 - \dfrac{s_i^{(j)} - s_i^{(j+1)}}{s_i^{(j)} - s_{i-1}^{(j+1)}}\right)}$$

with $s_0^{(j)} = r_{0,0}^{(j)}(z) = a_j$ and $s_{-1}^{(j)} = 0$ for $j = 0, 1, 2, \ldots$ and

$$b_i = \lim_{x \to z} r_{m,n}^{(0)}(x) = s_i^{(0)}$$

The values $s_i^{(j)}$ are ordered in a table as follows

$$
\begin{array}{ccccc}
s_0^{(0)} & & & & \\
& s_1^{(0)} & & & \\
s_0^{(1)} & & s_2^{(0)} & & \\
& s_1^{(1)} & & \ddots & \\
s_0^{(2)} & & s_2^{(1)} & & \\
& s_1^{(2)} & & \ddots & \\
s_0^{(3)} & & s_2^{(2)} & & \\
& s_1^{(3)} & \vdots & \ddots & \\
s_0^{(4)} & \vdots & & & \\
\vdots & & & &
\end{array}
$$

Let us compare the algorithm of Bulirsch-Stoer with the ϵ-algorithm, the qd-algorithm and the method of Neville-extrapolation based on polynomial interpolation (here $b_i = r_{i,0}^{(0)}(z)$). Input is the sequence

$$\{a_i\}_{i \in \mathbb{N}} = \{1/\sqrt{i+1}\}_{i \in \mathbb{N}}$$

with $\lim_{i \to \infty} a_i = 0$. For the methods based on interpolation we use

$$\{x_i\}_{i \in \mathbb{N}} = \{1/(i+1)\}_{i \in \mathbb{N}}$$

$$z = \lim_{i \to \infty} x_i = 0$$

The results are listed in table 4.9. Computations were performed in single precision arithmetic.

Table 4.9.

$$a_i = \frac{1}{\sqrt{i+1}}$$

i	a_i	Neville b_i	Bulirsch-Stoer $s_i^{(0)}$	epsilon $\epsilon_i^{(0)}$	qd $C_i(1)$
0	1.00000000	1.00000000	1.00000000	1.00000000	1.00000000
2	0.57735026	0.26964909	0.24118084	0.47414386	0.47414389
4	0.44721359	0.16024503	0.11929683	0.30954611	0.30954611
6	0.37796450	0.11422185	0.07439047	0.22960453	0.22960511
8	0.33333334	0.08890618	0.06251375	0.18284784	0.18285738
10	0.30151135	0.07302472	0.05196425	0.15728275	0.15730451
12	0.27735010	0.06171018	0.04694320	0.14115857	0.14109026
14	0.25819889	0.11071037	0.04458576	0.12546575	0.12571160
16	0.24253564	0.34646219	0.04921301	0.10151753	0.10491328
18	0.22941573	−2.86674523	0.04912448	0.10668034	0.10763902
20	0.21821788	−99.92565918	0.05485462	0.09624152	0.10445023

1.4. The ρ-algorithm.

Another technique is based on the construction of interpolating continued fractions of the form

$$d_0 + \left.\frac{x - x_0}{d_1}\right| + \left.\frac{x - x_1}{d_2}\right| + \left.\frac{x - x_2}{d_3}\right| + \ldots$$

If we take for $g_i(x)$ the ith convergent

$$d_0 + \sum_{k=1}^{i} \left.\frac{x - x_{k-1}}{d_k}\right|$$

and take $z = \infty$ as extrapolation point, then we can put for i even:

$$b_i = \lim_{x \to \infty} g_i(x) = d_0 + d_2 + \ldots + d_i$$

Now the sequence b_i can be calculated using reciprocal differences. Hence we can set up the following scheme:

$$t_{k,0} = a_k \qquad k = 0, 1, 2, \ldots$$

$$t_{k,1} = \frac{x_k - x_{k-1}}{a_k - a_{k-1}} \qquad k = 1, 2, 3, \ldots$$

$$t_{k,j} = t_{k-1,j-2} + \frac{x_k - x_{k-j}}{t_{k,j-1} - t_{k-1,j-1}} \qquad k = j, j+1, \ldots \text{ and } j = 2, 3, \ldots$$

where in fact, in the notation of chapter III,

$$t_{k,j} = \rho[x_{k-j}, \ldots, x_k]$$

Then clearly for i even

$$b_i = \rho[x_0, \ldots, x_i] = t_{i,i}$$

The values $t_{k,j}$ are ordered as in the following table

$$t_{0,0}$$

$$t_{1,0} \qquad t_{1,1}$$

$$t_{2,0} \qquad t_{2,1} \qquad t_{2,2}$$

$$t_{3,0} \qquad t_{3,1} \qquad t_{3,2} \qquad \ddots$$

$$\vdots \qquad \vdots \qquad \vdots$$

with $t_{k,0} = a_k$ and $t_{k,-1} = 0$ for $k = 0, 1, 2, \ldots$.
If we use $\{x_i\}_{i \in \mathbb{N}}$ as interpolation points for the algorithm of Bulirsch and Stoer
and $\{x_i'\}_{i \in \mathbb{N}}$ with

$$x_i' = \frac{1}{x_i}$$

as interpolation points for the ρ-algorithm, then for even i:

$$t_{i,i} = s_i^{(0)}$$

(see problem (3)).
Remark that the computation of $t_{i,i}$ takes a smaller effort than that of $s_i^{(0)}$. For
more properties of the ρ-algorithm we refer to [4]. We illustrate the influence
of the choice of the $\{x_i\}_{i \in \mathbb{N}}$ with some numerical examples. In the tables 4.10.
and 4.11. several $\{b_i\}_{i \in \mathbb{N}}$ were constructed, respectively for

$$\{a_i\}_{i \in \mathbb{N}} = \{1/\sqrt{i+1}\}_{i \in \mathbb{N}}$$

and

$$\{a_i\}_{i \in \mathbb{N}} = \{1/(i+1)\}_{i \in \mathbb{N}}$$

All computations were performed in single precision arithmetic. In each table
the first sequence $\{b_i\}_{i \in \mathbb{N}}$ is constructed with $x_i = \sqrt{i}$, the second with $x_i = i$,
the third with $x_i = i^2$ and the fourth with $x_i = e^i$. The choice of the x_i greatly
influences the convergence behaviour of the resulting sequence $\{b_i\}_{i \in \mathbb{N}}$. We see
that we do not necessarily obtain better results when the x_i converge faster to
infinity. More information on this matter can be found in [6].

Table 4.10.

$$a_i = \frac{1}{\sqrt{i+1}}$$

i	a_i	$b_i = t_{i,i}$ $x_i = \sqrt{i}$	$b_i = t_{i,i}$ $x_i = i$	$b_i = t_{i,i}$ $x_i = i^2$	$b_i = t_{i,i}$ $x_i = e^i$
0	1.00000000	1.00000000	1.00000000	1.00000000	1.00000000
2	0.57735026	7.07811832	0.24118099	0.50412309	0.49505624
4	0.44721359	0.02922609	0.11929671	0.30890498	0.39424682
6	0.37796450	−0.00126043	0.07431557	0.22015348	0.34179661
8	0.33333334	−0.00124686	0.06307473	0.17057094	0.30704349
10	0.30151135	−0.00416392	0.05177462	0.14054570	0.28143203
12	0.27735010	0.00019596	0.04759169	0.12608825	0.26141584
14	0.25819889	−0.00047069	0.04435266	0.11451113	0.24517243
16	0.24253564	0.00017922	0.04089086	0.10530391	0.23163427
18	0.22941573	0.00026065	0.05994635	0.09724047	0.22012024
20	0.21821788	0.00000440	0.03507820	0.09777263	0.21016997

Table 4.11.

$$a_i = \frac{1}{i+1}$$

i	a_i	$b_i = t_{i,i}$ $x_i = \sqrt{i}$	$b_i = t_{i,i}$ $x_i = i$	$b_i = t_{i,i}$ $x_i = i^2$	$b_i = t_{i,i}$ $x_i = e^i$
0	1.00000000	1.00000000	1.00000000	1.00000000	1.00000000
2	0.33333334	−2.41421294	0.00000006	0.25000003	0.24015641
4	0.20000000	0.00000162	0.00000003	0.09090916	0.15439278
6	0.14285715	0.00000038	0.00000002	0.04545459	0.11652596
8	0.11111111	−0.00000042	−0.00000015	0.02703313	0.09417003
10	0.09090909	0.00000018	0.00000003	0.01793462	0.07915991

In [49] a study was made of some nonlinear convergence accelerating methods. The ε-algorithm was found to be the most effective one in many cases. Therefore we now generalize the ε-algorithm for the convergence acceleration of multi-dimensional tables.

1.5. The multivariate ε-algorithm.

A sequence $\{a_i\}_{i \in \mathbb{N}}$ can be considered as a table with single entry. For its convergence acceleration we constructed the univariate function

$$f(x) = \sum_{i=0}^{\infty} \nabla a_i x^i$$

with $a_i = 0$ for $i < 0$ and calculated diagonal Padé approximants evaluated at $x = 1$. Let us now first consider a table

$$\{a_{j,k}\}_{j,k \in \mathbb{N}}$$

with double entry and with $A = \lim_{j,k \to \infty} a_{j,k} = \lim_{j \to \infty}(\lim_{k \to \infty} a_{jk}) = \lim_{k \to \infty}(\lim_{j \to \infty} a_{jk})$. To accelerate its convergence we introduce

$$f(x,y) = \sum_{j,k=0}^{\infty} \nabla a_{jk} x^j y^k$$

with $\nabla a_{jk} = a_{jk} - a_{j,k-1} - a_{j-1,k} + a_{j-1,k-1}$ and $a_{jk} = 0$ for $j < 0$ or $k < 0$. Clearly

$$f(1,1) = \lim_{j,k \to \infty} a_{jk}$$

We will again construct a sequence $\{b_i\}_{i \in \mathbb{N}}$ that will converge faster to A in some cases [11]. To this end we now calculate bivariate Padé approximants for $f(x,y)$ and evaluate them at $(x,y) = (1,1)$.
If we denote by

$$\sigma_\ell = \sum_{j+k=\ell} a_{jk} = a_{\ell,0} + a_{\ell-1,1} + \ldots + a_{1,\ell-1} + a_{0,\ell}$$

and start the ε-algorithm with the partial sums of $f(1,1)$

$$\epsilon_0^{(\ell)} = \sum_{j+k=0}^{\ell} \nabla a_{j,k} = \sigma_\ell - \sigma_{\ell-1} \qquad \ell = 0, 1, \ldots$$

then the bivariate diagonal Padé approximants $r_{i,i}(x, y)$ evaluated at $(x, y) = (1, 1)$ are given by $\epsilon_{2i}^{(0)}$.

Hence we put

$$b_i = \epsilon_{2i}^{(0)}$$

Let us generalize the idea for a table with multiple entry $\{a_{j_1 \ldots j_k}\}_{j_1, \ldots, j_k \in \mathbb{N}}$. We define

$$f(x_1, \ldots, x_k) = \sum_{j_1, \ldots, j_k = 0}^{\infty} \nabla a_{j_1 \ldots j_k} \, x_1^{j_1} \ldots x_k^{j_k}$$

with

$$\nabla a_{j_1 \ldots j_k} = a_{j_1 \ldots j_k} - \sum_{\ell=1}^{k} a_{j_1 \ldots (j_\ell - 1) \ldots j_k}$$
$$+ \sum_{\substack{\ell, m = 1 \\ \ell < m}}^{k} a_{j_1 \ldots j_{\ell-1}(j_\ell - 1)j_{\ell+1} \ldots j_{m-1}(j_m - 1)j_{m+1} \ldots j_k}$$
$$- \ldots + (-1)^k \, a_{(j_1 - 1) \ldots (j_k - 1)}$$

It is easy to prove that

$$f(1, 1, \ldots, 1) = \lim_{j_1, \ldots, j_k \to \infty} a_{j_1 \ldots j_k}$$

Again multivariate Padé approximants for $f(x_1, \ldots, x_k)$ can be calculated and evaluated at $(x_1, \ldots, x_k) = (1, \ldots, 1)$ via the ε-algorithm.

Since

$$\sum_{j_1 + \ldots + j_k = \ell} \nabla a_{j_1 \ldots j_k} = \sum_{j=0}^{k} (-1)^j \binom{k}{j} \sigma_{\ell - j}$$

where now

$$\sigma_\ell = \sum_{j_1 + \ldots + j_k = \ell} a_{j_1 \ldots j_k}$$

the $\epsilon_0^{(\ell)}$ for $f(x_1, \ldots, x_k)$ are given by

$$\epsilon_0^{(\ell)} = \sum_{j_1 + \ldots + j_k = 0}^{\ell} \nabla a_{j_1 \ldots j_k} = \sum_{j=0}^{k-1} (-1)^j \binom{k-1}{j} \sigma_{\ell - j}$$

We illustrate this type of convergence acceleration with a numerical example. Suppose one wants to calculate the integral of a function $G(x_1, \ldots, x_k)$ on a bounded closed domain D of $\mathbf{R}^k$. Let $D = [0,1] \times \ldots \times [0,1]$ for the sake of simplicity. The table $\{a_{j_1 \ldots j_k}\}_{j_1, \ldots, j_k \in \mathbb{N}}$ can be obtained for instance by subdividing the interval $[0,1]$ in the ℓth direction $(\ell = 1, \ldots, k)$ into 2^{j_ℓ} intervals of equal length $h_\ell = 2^{-j_\ell}(j_\ell = 0, 1, 2, \ldots)$. Using the midpoint-rule one can then substitute approximations

$$\int_0^{h_1} \ldots \int_0^{h_k} G(x_1, \ldots, x_k)dx_1 \ldots dx_k = h_1 \ldots h_k G\left(\frac{h_1}{2}, \ldots, \frac{h_k}{2}\right)$$

to calculate the $a_{j_1 \ldots j_k}$.

We restrict ourselves again to the case of two variables. With $h_1 = 2^{-j}$ and $h_2 = 2^{-k}$ we get

$$a_{jk} = \frac{1}{2^{j+k}}\left(\sum_{\ell=1}^{2^j} \sum_{m=1}^{2^k} G\left(\frac{2\ell - 1}{2^{j+1}}, \frac{2m - 1}{2^{k+1}}\right)\right)$$

The column $\epsilon_0^{(\ell)}(\ell = 0, 1, \ldots)$ in the ϵ-table given by

$$\epsilon_0^{(\ell)} = \sum_{j=0}^{k-1} (-1)^j \binom{k-1}{j} \sigma_{\ell-j}$$

was also used by Genz [21] to start the ϵ-algorithm for the approximate calculation of multidimensional integrals by means of extrapolation methods. He preferred this method to six other methods because of its simplicity and general use of fewer integrand evaluations.

For

$$G(x, y) = \frac{1}{x + y}$$

we have

$$\lim_{j,k \to \infty} a_{jk} = \int_0^1 \int_0^1 \frac{1}{x + y}dxdy = 2\ln 2 = 1.386294361119891\ldots$$

In table 4.12. one can find the a_{ii} slowly converging to the exact value of the integral because of the singularity of the integrand in the origin. The $b_i = r_{i,i}(1, 1)$ converge much faster. For the calculation of the b_i we need the σ_ℓ $(\ell = 0, \ldots, 2i)$.

Hence b_i should be compared with an a_{jk} which uses the same amount of information, i.e. $j + k = 2i$.

Table 4.12.

$$\int_0^1 \int_0^1 \frac{1}{x+y}\,dxdy = 1.386294361119891\ldots$$

i	a_{ii}	$b_i = \epsilon_{2i}^{(0)}$
1	1.166666666667	1.330294906166
2	1.269047619048	1.396395820203
3	1.325743700744	1.386872037696
4	1.355532404415	1.386308917778

§2. Nonlinear equations.

Suppose we want to find a root x^* of the nonlinear equation

$$f(x) = 0$$

Here the function f may be real- or complex-valued. If f is now replaced by a local approximation then a zero of that local approximation could be considered as an approximation for x^*. Methods based on this reasoning are called **direct**. One could also consider the inverse function g of f in a neighbourhood of the origin, if it exists, and replace g by a local approximation. Then an evaluation of this local approximation in 0 could be considered as an approximation for x^* since

$$g(0) = x^*$$

Methods using this technique are called **inverse**.

2.1. Iterative methods based on Padé approximation.

Let x_i be an approximation for the root x^* of f and let

$$r_i(x) = \frac{p_i}{q_i}(x)$$

be the Padé approximant of order (m, n) for f in x_i. Then the next approximation x_{i+1} is calculated such that

$$p_i(x_{i+1}) = 0$$

In case $p_i(x)$ is linear $(m = 1)$ the value x_{i+1} is uniquely determined. It is clear that this is to be preferred for the sake of simplicity. A well-known method obtained in this way is Newton's method $(m = 1, n = 0)$ which can be derived as follows. The Taylor series expansion for $f(x)$ at x_i is given by

$$f(x) = f(x_i) + f'(x_i)(x - x_i) + \frac{f''(x_i)}{2}(x - x_i)^2 + \dots \tag{4.2.}$$

Hence the Padé approximant of order $(1, 0)$ for f at x_i equals

$$r_i(x) = f(x_i) + f'(x_i)(x - x_i)$$

and we obtain

$$x_{i+1} = x_i - \frac{f(x_i)}{f'(x_i)} \tag{4.3.}$$

Another famous method is Halley's method based on the use of the Padé approximant of order $(1, 1)$ for f at x_i:

$$x_{i+1} = x_i - \frac{f(x_i)/f'(x_i)}{1 - \frac{1}{2}f''(x_i)\dfrac{f(x_i)}{f'(x_i)^2}} \tag{4.4.}$$

Since the iterative procedures (4.3.) and (4.4.) only use information in the point x_i to calculate the next iteration point x_{i+1} they are called **one-point**. The order of convergence of these iterative methods is given in the next theorem.

Theorem 4.4.

If $\{x_i\}_{i \in \mathbb{N}}$ converges to a simple root x^* of f and if $r_i(x)$ is normal for every i then

$$\lim_{i \to \infty} \frac{|x_{i+1} - x^*|}{|x_i - x^*|^{m+n+1}} = C^* < \infty$$

For the proof we refer to [52]. Similar results were given in [18], [43] and [44]. So the order is at least $m + n + 1$ and it depends only on the sum of m and n. Consequently the order is not changed when Padé approximants lying on an ascending diagonal in the Padé table are used. Iterative methods resulting from the use of (m, n) Padé approximants with $n > 0$ can be interesting because the **asymptotic error constant** C^* may be smaller than when $n = 0$ [43].
The iterative procedures (4.3) and (4.4) can also be derived as inverse methods (see problem (5)).
Let us apply Newton's and Halley's method to the solution of

$$f(x) = \frac{\sin x}{x - 0.1} = 0$$

The root $x^* = 0$. We use $x_0 = 0.09$ as initial point. The next iteration steps can be found in table 4.13. Computations were performed in double precision accuracy (56–digit binary arithmetic).

Table 4.13.

i	Newton x_i	Halley x_i
0	0.90000000D—01	0.90000000D—01
1	0.80997568D—01	—0.40647645D—02
2	0.65599659D—01	0.35821646D—06
3	0.43022008D—01	—0.24514986D—18
4	0.18500311D—01	0.00000000D+00
5	0.34212128D—02	
6	0.11703452D—03	
7	0.13697026D—06	
8	0.18760851D—12	
9	0.35197080D—24	
10	0.00000000D+00	

It is obvious that a method based on the use of (m, n) Padé approximants for f with $n > 0$ gives better results here: the function f has a singularity at $x = 0.1$. Observe that in the Newton-iteration x_6 is a good initial point in the sense that from there on quadratic convergence is guaranteed:

$$|x_{i+1} - x^*| = |x_{i+1}| \simeq C^*|x_i - x^*|^2 \text{ for } i \geq 0$$

with $C^* = 10$. For the Halley iteration we clearly have cubic convergence from x_1 on.

The formulas (4.3.) and (4.4.) can also be generalized for the solution of a system of nonlinear equations

$$\begin{cases} f_1(x_1, \ldots, x_k) = 0 \\ \vdots \\ f_k(x_1, \ldots, x_k) = 0 \end{cases}$$

which we shall write as

$$F(x_1, \ldots, x_k) = 0$$

Newton's method can then be expressed as [45]

$$
\begin{pmatrix} x_1^{(i+1)} \\ \vdots \\ x_k^{(i+1)} \end{pmatrix} = \begin{pmatrix} x_1^{(i)} \\ \vdots \\ x_k^{(i)} \end{pmatrix} - F'(x_1^{(i)}, \ldots, x_k^{(i)})^{-1} \begin{pmatrix} f_1(x_1^{(i)}, \ldots, x_k^{(i)}) \\ \vdots \\ f_k(x_1^{(i)}, \ldots, x_k^{(i)}) \end{pmatrix} \qquad (4.5.)
$$

where $F'(x_1^{(i)}, \ldots, x_k^{(i)})$ is the Jacobian matrix of first partial derivatives evaluated at $(x_1^{(i)}, \ldots, x_k^{(i)})$ with

$$
F'(x_1, \ldots, x_k) = \begin{pmatrix} \dfrac{\partial f_1}{\partial x_1} & \cdots & \dfrac{\partial f_1}{\partial x_k} \\ \vdots & & \vdots \\ \dfrac{\partial f_k}{\partial x_1} & \cdots & \dfrac{\partial f_k}{\partial x_k} \end{pmatrix}
$$

Let us now introduce the abbreviations

$$
F_i = F(x_1^{(i)}, \ldots, x_k^{(i)})
$$
$$
F_i' = F'(x_1^{(i)}, \ldots, x_k^{(i)})
$$

To generalize Halley's method we first rewrite (4.4.) as

$$
x_{i+1} = x_i + \frac{(-f(x_i)/f'(x_i))^2}{\dfrac{-f(x_i)}{f'(x_i)} + \dfrac{f''(x_i)f(x_i)^2}{2f'(x_i)^3}}
$$

Then for the solution of a system of equations it becomes [14]

$$
\begin{pmatrix} x_1^{(i+1)} \\ \vdots \\ x_k^{(i+1)} \end{pmatrix} = \begin{pmatrix} x_1^{(i)} \\ \vdots \\ x_k^{(i)} \end{pmatrix} +
$$

$$
+ \frac{(-F_i'^{-1}F_i)^2}{-F_i'^{-1}F_i + \dfrac{1}{2}F_i'^{-1}F''(x_1^{(i)}, \ldots, x_k^{(i)})[-F_i'^{-1}F_i, -F_i'^{-1}F_i]} \qquad (4.6.)
$$

where the division and the square are performed componentwise and $F''(x_1, \ldots, x_k)$ is the hypermatrix of second partial derivatives given by

$$F''(x_1, \ldots, x_k) =$$

$$\begin{pmatrix} \dfrac{\partial^2 f_1}{\partial x_1^2} \cdots \dfrac{\partial^2 f_1}{\partial x_k \partial x_1} & \dfrac{\partial^2 f_1}{\partial x_1 \partial x_2} \cdots \dfrac{\partial^2 f_1}{\partial x_k \partial x_2} & \cdots & \dfrac{\partial^2 f_1}{\partial x_1 \partial x_k} \cdots \dfrac{\partial^2 f_1}{\partial x_k^2} \\ \vdots & \vdots \quad \vdots & & \vdots & \vdots \\ \dfrac{\partial^2 f_k}{\partial x_1^2} \cdots \dfrac{\partial^2 f_k}{\partial x_k \partial x_1} & \dfrac{\partial^2 f_k}{\partial x_1 \partial x_2} \cdots \dfrac{\partial^2 f_k}{\partial x_k \partial x_2} & \cdots & \dfrac{\partial^2 f_k}{\partial x_1 \partial x_k} \cdots \dfrac{\partial^2 f_k}{\partial x_k^2} \end{pmatrix}$$

which we have to multiply twice with the vector $-F_i'^{-1} F_i$. This multiplication is performed as follows. The hypermatrix $F''(x_1, \ldots, x_k)$ is a row of k matrices, each $k \times k$. If we use the usual matrix-vector multiplication for each element in the row we obtain

$$F''(x_1, \ldots, x_k) \begin{pmatrix} y_1 \\ \vdots \\ y_k \end{pmatrix} \begin{pmatrix} y_1 \\ \vdots \\ y_k \end{pmatrix} = \begin{pmatrix} \sum\limits_{i=1}^{k} \dfrac{\partial^2 f_1}{\partial x_1 \partial x_i} y_i \cdots \sum\limits_{i=1}^{k} \dfrac{\partial^2 f_1}{\partial x_k \partial x_i} y_i \\ \vdots \qquad \vdots \\ \sum\limits_{i=1}^{k} \dfrac{\partial^2 f_k}{\partial x_1 \partial x_i} y_i \cdots \sum\limits_{i=1}^{k} \dfrac{\partial^2 f_k}{\partial x_k \partial x_i} y_i \end{pmatrix} \begin{pmatrix} y_1 \\ \vdots \\ y_k \end{pmatrix}$$

In [14] is proved that the iterative procedure (4.6.) actually results from the use of multivariate Padé approximants of order (1,1) for the inverse operator of $F(x_1, \ldots, x_k)$ at $(x_1^{(i)}, \ldots, x_k^{(i)})$.

To illustrate the use of the formulas (4.5.) and (4.6.) we shall now solve the nonlinear system

$$\begin{cases} f_1(x,y) = e^{-x+y} - 0.1 = 0 \\ f_2(x,y) = e^{-x-y} - 0.1 = 0 \end{cases}$$

which has a simple root at

$$\begin{pmatrix} -\ln(0.1) \\ 0. \end{pmatrix} = \begin{pmatrix} 2.302585092994046... \\ 0. \end{pmatrix}$$

As initial point we take $(x^{(0)}, y^{(0)}) = (4.3, 2.0)$. In table 4.14. one finds the consecutive iteration steps of Newton's and Halley's method. Again Halley's method behaves much better than the polynomial method of Newton. Here the inverse operator G of the system of equations F has a singularity near to the origin and this singularity causes trouble if we get close to it. For

$$F(x,y) = \begin{pmatrix} f_1(x,y) \\ f_2(x,y) \end{pmatrix} = \begin{pmatrix} u \\ v \end{pmatrix}$$

we can write

$$G(u,v) = \begin{pmatrix} g_1(u,v) \\ g_2(u,v) \end{pmatrix} = \begin{pmatrix} -0.5(\ln(0.1+u) + \ln(0.1+v)) \\ 0.5(\ln(0.1+u) - \ln(0.1+v)) \end{pmatrix} = \begin{pmatrix} x \\ y \end{pmatrix}$$

With $(x^{(0)}, y^{(0)}) = (4.3, 2.0)$ the value $v^{(0)} = f_2(x^{(0)}, y^{(0)})$ is close to -0.1 which is close to the singularity of G.

For the computation of the Padé approximants involved in all these methods the ϵ-algorithm can be used. Another iterative procedure for the solution of a system of nonlinear equations based on the ϵ-algorithm but without the evaluation of derivatives can be found in [5]. Since it does not result from approximating the multivariate nonlinear problem by a multivariate rational function, we do not discuss it here.

Table 4.14.

i	Newton $x^{(i)}$	Newton $y^{(i)}$	Halley $x^{(i)}$	Halley $y^{(i)}$
0	0.43000000D+01	0.20000000D+01	0.43000000D+01	0.20000000D+01
1	—0.22447305D+02	—0.24729886D+02	0.28798400D+01	0.57957286D+00
2	—0.21927303D+02	—0.24229888D+02	0.22495475D+01	—0.52625816D—01
3	—0.21427303D+02	—0.23729888D+02	0.23018229D+01	—0.44901947D—02
4	—0.20927303D+02	—0.23229888D+02	0.23025841D+01	—0.57154737D—05
5	—0.20427303D+02	—0.22729888D+02	0.23025851D+01	—0.97689305D—11
6	—0.19927303D+02	—0.22229888D+02	0.23025851D+01	—0.17438130D—16
7	—0.19427303D+02	—0.21729888D+02		
8	—0.18927303D+02	—0.21229888D+02		
9	0.18427303D+02	—0.20729888D+02		
10	—0.17927303D+02	—0.20229888D+02		
11	—0.17427303D+02	—0.19729888D+02		
12	—0.16927303D+02	—0.19229888D+02		
13	—0.16427303D+02	—0.18729888D+02		
14	—0.15927303D+02	—0.18229888D+02		
15	—0.15427303D+02	—0.17729888D+02		
16	—0.14927303D+02	—0.17229888D+02		
17	—0.14427303D+02	—0.16729888D+02		
18	—0.13927303D+02	—0.16229888D+02		
19	—0.13427303D+02	—0.15729888D+02		
20	—0.12927303D+02	—0.15229888D+02		

2.2. Iterative methods based on rational interpolation.

Let

$$r_i(x) = \frac{p_i}{q_i}(x)$$

with p_i and q_i respectively of degree m and n, be such that in an approximation x_i for the root x^* of f

$$
\begin{aligned}
r_i^{(\ell)}(x_i) &= f^{(\ell)}(x_i) & \ell &= 0, \ldots, s_0 - 1 \\
r_i^{(\ell)}(x_{i-1}) &= f^{(\ell)}(x_{i-1}) & \ell &= 0, \ldots, s_1 - 1 \\
&\;\;\vdots & & \\
r_i^{(\ell)}(x_{i-j}) &= f^{(\ell)}(x_{i-j}) & \ell &= 0, \ldots, s_j - 1
\end{aligned}
\tag{4.7.}
$$

with $m + n + 1 = \sum_{\ell=0}^{j} s_\ell$. Then the next iteration step x_{i+1} is computed such that

$$p_i(x_{i+1}) = 0$$

For the calculation of x_{i+1} we now use information in more than one previous point. Hence such methods are called **multipoint**. Their order of convergence can be calculated as follows.

Theorem 4.5.

If $\{x_i\}_{i \in \mathbb{N}}$ converges to a simple root x^* of f and $f^{(m+n+1)}(x)$ with $n > 0$ is continuous in a neighbourhood of x^* with

$$
\begin{vmatrix}
f^{(m)}(x^*) & f^{(m-1)}(x^*) & \ldots & f^{(m-n+1)}(x^*) \\
f^{(m+1)}(x^*) & f^{(m)}(x^*) & \ldots & f^{(m-n+2)}(x^*) \\
\vdots & \vdots & \ddots & \vdots \\
f^{(m+n-1)}(x^*) & f^{(m+n-2)}(x^*) & \ldots & f^{(m)}(x^*)
\end{vmatrix} \neq 0
$$

where $f^{(k)}(x^*) = 0$ if $k < 0$, then the order of the iterative method based on the use of $r_i(x)$ satisfying (4.7.) is the unique positive root of the polynomial

$$x^{j+1} - s_0 x^j - s_1 x^{j-1} - \ldots - s_j = 0$$

The proof can be found in [52].

If we restrict ourselves to the case

$$s_\ell = s \qquad \ell = 0, \ldots, j$$

then it is interesting to note that the unique positive root of

$$x^{j+1} - s \sum_{\ell=0}^{j} x^\ell$$

increases with j but is bounded above by $s + 1$ [53 pp. 46-52]. As a conclusion we may say that the use of large j is not recommendable. We give some examples. Take $m = 1$, $n = 1$, $s = 1$ and $j = 2$. Then x_{i+1} is given by

$$x_{i+1} = x_i - \frac{f(x_i)[f(x_{i-1}) - f(x_{i-2})]}{f(x_{i-1})\dfrac{f(x_{i-2}) - f(x_i)}{x_{i-2} - x_i} - f(x_{i-2})\dfrac{f(x_{i-1}) - f(x_i)}{x_{i-1} - x_i}} \qquad (4.8.)$$

The order of this method is 1.84, which is already very close to $s + 1 = 2$. Take $m = 1$, $n = 1$, $s_0 = 2$, $s_1 = 1$ and $j = 1$. Then x_{i+1} is given by

$$x_{i+1} = x_i + \frac{f(x_i)(x_i - x_{i-1})}{f(x_{i-1})\ f'(x_i)\dfrac{(x_i - x_{i-1})}{f(x_i) - f(x_{i-1})} - f(x_i)} \qquad (4.9.)$$

The order of this procedure is 2.41. The case $m = 1$, $n = 0$, $s = 1$ and $j = 1$ reduces to the secant method with order 1.62.

Let us again calculate the root of

$$f(x) = \frac{\sin x}{x - 0.1} = 0$$

with initial points close to the singularity in $x = 0.1$. The successive iteration steps computed in double precision (56–digit binary arithmetic) are shown in table 4.15.

Table 4.15.

i	(4.8.) x_i	(4.9.) x_i	secant method x_i
	0.80000000D—01		
	0.90000000D—01	0.80000000D—01	0.80000000D—01
0	0.85000000D—01	0.90000000D—01	0.90000000D—01
1	—0.15847802D—03	— 0.17324153D—03	0.71995917D—01
2	0.15131452D—06	0.46621186D—10	0.64791421D—01
3	—0.51538368D—13	—0.61392311D—25	0.46636684D—01
4	0.20630843D—24	0.00000000D+00	0.30206020D—01
5	0.00000000D+00		0.14080364D—01
6			0.42512345D—02
7			0.59843882D—03
8			0.25439081D—04
9			0.15223576D—06
10			0.38727362D—10
11			0.58956894D—16
12			0.22832450D—25

By means of the multivariate Newton-Padé approximants introduced in section 6.2. of chapter III the previous formulas can be generalized for the solution of systems of nonlinear equations. We use the same notations as in chapter III and as in the previous section. For each of the multivariate functions $f_j(x_1, \ldots, x_k)$ with $j = 1, \ldots, k$ we choose

$$D = N = \{(0, \ldots, 0), (1, 0, \ldots, 0), (0, 1, 0, \ldots, 0), \ldots, (0, \ldots, 0, 1)\} \subseteq \mathbb{N}^k$$

$$H = \{(2, 0, \ldots, 0), (0, 2, 0, \ldots, 0), \ldots, (0, \ldots, 0, 2)\} \subseteq \mathbb{N}^k$$

Here the interpolationset $N \cup H$ expresses interpolation conditions in the points

$$\left(x_1^{(i)}, \ldots, x_k^{(i)}\right), \left(x_1^{(i-1)}, x_2^{(i)}, \ldots, x_k^{(i)}\right), \ldots, \left(x_1^{(i)}, \ldots, x_{k-1}^{(i)}, x_k^{(i-1)}\right),$$

$$\left(x_1^{(i-2)}, x_2^{(i)}, \ldots, x_k^{(i)}\right), \ldots, \left(x_1^{(i)}, \ldots, x_{k-1}^{(i)}, x_k^{(i-2)}\right)$$

Remark that this set of interpolation points is constructed from only three successive iteration points.
The numerator of

$$r_{i_j}(x_1, \ldots, x_k) = \frac{p_{i_j}}{q_{i_j}}(x_1, \ldots, x_k) \qquad j = 1, \ldots, k$$

satisfying

$$(f_j q_{i_j} - p_{i_j})(x_1, \ldots, x_k) = \sum_{(\ell_1, \ldots, \ell_k) \in \mathbf{N}^k \backslash (N \cup H)} d_{\ell_1 \ldots \ell_k} \; B_{\ell_1 \ldots \ell_k}(x_1, \ldots, x_k)$$

where

$$B_{\ell_1 \ldots \ell_k}(x_1, \ldots, x_k) = \prod_{\ell=0}^{\ell_1 - 1} (x_1 - x_1^{(i-\ell)}) \ldots \prod_{\ell=0}^{\ell_k - 1} (x_k - x_k^{(i-\ell)})$$

with possible coalescence of points, is then given by

$$p_{i_j}(x_1, \ldots, x_k) =$$

$$\begin{vmatrix} N_{0,\ldots,0}(x_1, \ldots, x_k) & N_{1,0,\ldots,0}(x_1, \ldots, x_k) & \ldots & N_{0,\ldots,0,1}(x_1, \ldots, x_k) \\ c_{02,00,\ldots,00} & c_{12,00,\ldots,00} & \ldots & 0 \\ \vdots & & \ddots & \\ c_{00,\ldots,00,02} & 0 & \ldots & c_{00,\ldots,00,12} \end{vmatrix}$$

where

$$N_{i_1,\ldots,i_k}(x_1, \ldots, x_k) = \sum_{(\ell_1,\ldots,\ell_k) \in N} c_{i_1 \ell_1, \ldots, i_k \ell_k} B_{\ell_1 \ldots \ell_k}(x_1, \ldots, x_k)$$

The values $c_{s_1 t_1, \ldots, s_k t_k}$ are multivariate divided differences with possible coalescence of points. Remark that this formula is only valid if the set H provides a system of linearly independent equations. The next iterationstep $(x_1^{(i+1)}, \ldots, x_k^{(i+1)})$ is then constructed such that

$$\begin{cases} p_{i_1}(x_1^{(i+1)}, \ldots, x_k^{(i+1)}) = 0 \\ \vdots \\ p_{i_k}(x_1^{(i+1)}, \ldots, x_k^{(i+1)}) = 0 \end{cases}$$

For $k = 1$ and without coalescence of points this procedure coincides with the iterative method (4.8.).

With $k = 2$ and without coalescence of points we obtain a bivariate generalization of (4.8.).

Let us use this technique to solve the system

$$\begin{cases} e^{-x+y} = 0.1 \\ e^{-x-y} = 0.1 \end{cases}$$

with initial points $(3.2, -0.95)$, $(3.4, -1.15)$ and $(3.3, -1.00)$. The numerical results computed in double precision (56–digit binary arithmetic) are displayed in table 4.16. The simple root is $(2.302585092994046 \ldots, 0.)$.

In this way we can also derive a discretized Newton method in which the partial derivatives of the Jacobian matrix are approximated by difference quotients

$$\begin{aligned} N =& H = \{(0, \ldots, 0), (1, 0, \ldots, 0), \ldots, (0, \ldots, 0, 1)\} \\ & D = \{(0, \ldots, 0)\} \end{aligned}$$

$$\left. \frac{\partial f_j}{\partial x_\ell} \right|_{(x_1^{(i)}, \ldots, x_k^{(i)})} \simeq \frac{f_j(x_1^{(i)}, \ldots, x_{\ell-1}^{(i)}, x_\ell^{(i-1)}, x_{\ell+1}^{(i)}, \ldots, x_k^{(i)}) - f_j(x_1^{(i)}, \ldots, x_k^{(i)})}{x_\ell^{(i-1)} - x_\ell^{(i)}}$$

If we call this matrix of difference quotients ΔF_i, then the next iterate is computed by means of

$$\begin{pmatrix} x_1^{(i+1)} \\ \vdots \\ x_k^{(i+1)} \end{pmatrix} = \begin{pmatrix} x_1^{(i)} \\ \vdots \\ x_k^{(i)} \end{pmatrix} - (\Delta F_i)^{-1} \begin{pmatrix} f_1(x_1^{(i)}, \ldots, x_k^{(i)}) \\ \vdots \\ f_k(x_1^{(i)}, \ldots, x_k^{(i)}) \end{pmatrix}$$

As an example we take the same system of equations and the same but fewer initial points as above. The consecutive iteration steps computed in double precision (56–digit binary arithmetic) can now be found in table 4.17.

Table 4.16.

i	$x^{(i)}$	$y^{(i)}$
	0.32000000D+01	−0.95000000D+00
	0.34000000D+01	−0.11500000D+01
0	0.33000000D+01	−0.10000000D+01
1	0.25249070D+01	−0.22072875D+00
2	0.22618832D+01	0.41971944D−01
3	0.23127609D+01	−0.10164490D−01
4	0.23030978D+01	−0.51269373D−03
5	0.23025801D+01	0.49675854D−05
6	0.23025851D+01	−0.25696929D−08
7	0.23025851D+01	−0.12778916D−13
8	0.23025851D+01	−0.11350932D−16

Table 4.17.

i	$x^{(i)}$	$y^{(i)}$
	0.34000000D+01	−0.11500000D+01
0	0.33000000D+01	−0.10000000D+01
1	−0.29618530D+00	0.21743633D+01
2	0.32743183D+01	0.20884933D+01
3	0.22114211D+01	−0.84011352D+01
4	0.36513339D+01	−0.72149651D+01
5	−0.17900983D+04	0.20854111D+04

divergence

The rational method is again giving better results. Now the initial points are such that $u = f_1(x, y)$ is close to -0.1 which is precisely a singularity of the inverse operator for the considered system of nonlinear equations.

For a more stable variant of the discretized Newton method we refer to [24].

2.3. Iterative methods using continued fractions.

If $r_i(x)$ is the rational interpolant of order $(m, 1)$ for $\frac{1}{f}(x)$, satisfying

$$r_i(x^{(i-\ell)}) = \frac{1}{f}(x^{(i-\ell)}) \qquad \ell = 0, \ldots, m+1$$

then $r_i(x)$ can be written in the form

$$r_i(x) = d_0 + d_1(x - x^{(i-m-1)}) + \ldots + d_{m-1}(x - x^{(i-m-1)})\ldots(x - x^{(i-3)})$$
$$+ \frac{d_m(x - x^{(i-m-1)})\ldots(x - x^{(i-2)})}{1} \bigg| + \frac{x - x^{(i-1)}}{d_{m+1}} \bigg|$$

The coefficients $d_j (j = 0, 1, \ldots, m)$ are divided differences while d_{m+1} is an inverse difference. The root of $\frac{1}{r_i}(x)$ can be considered as an approximation for the root x^* of f. So

$$x^{(i+1)} = x^{(i-1)} - d_{m+1}$$

This method can be compared with methods based on the use of rational interpolants of order $(1, m)$ for $f(x)$.

2.4. The qd-algorithm.

First we state an important analytical property of the qd-scheme. To this end we introduce the following notation. For the function $f(x)$ given by its Taylor series development

$$f(x) = c_0 + c_1 x + c_2 x^2 + \ldots$$

we define the Hankel-determinants

$$H_{m,n+1} = \begin{vmatrix} c_m & c_{m+1} & \cdots & c_{m+n} \\ c_{m+1} & c_{m+2} & \cdots & c_{m+n+1} \\ \vdots & & & \vdots \\ c_{m+n} & c_{m+n+1} & \cdots & c_{m+2n} \end{vmatrix}$$

with $H_{m,0} = 1$.

We now call the series $f(x)$ **ultimately k-normal** if for every n with $0 \le n \le k$ there exists an integer M_n such that for $m \ge M_n$ the determinant $H_{m,n}$ is nonzero.

Theorem 4.6.

Let the meromorphic function $f(x)$ be given by its Taylor series development in the disk $B(0, r) = \{x \in \mathbb{C} \mid |x| < r\}$ and let the poles w_i of f in $B(0, r)$ be numbered such that

$$0 < |w_1| \le |w_2| \le \ldots < r$$

each pole occuring as many times as indicated by its order. If the series $f(x)$ is ultimately k-normal for some integer $k > 0$, then the qd-scheme associated with this series has the following properties:

a) for each n such that $0 < n \le k$ and such that $|w_{n-1}| < |w_n| < |w_{n+1}|$, where $w_0 = 0$ and, if f has exactly k poles, $w_{k+1} = \infty$, we have

$$\lim_{m \to \infty} q_n^{(m)} = \frac{1}{w_n}$$

b) for each n such that $0 < n \le k$ and such that $|w_n| < |w_{n+1}|$, we have

$$\lim_{m \to \infty} e_n^{(m)} = 0$$

The proof can be found in [29 pp. 612-613].

As a consequence of this theorem the qd-algorithm is an ingenious tool to determine the zeros of polynomials and entire functions. For if $p(x)$ is a polynomial then the zeros of $p(x)$ are the poles of the meromorphic function $f(x) = \frac{1}{p}(x)$. Any q-column corresponding to a simple pole of isolated modulus would tend to the reciprocal value of that pole. It would be flanked by e-columns that tend to zero. If moreover $f(x)$ is rational the last e-column would be zero. Unfortunately the qd-scheme as generated in section 3 of chapter II is numerically unstable. Rounding errors play an important role due to the divisions by small e-values. However, the rhombus rules (2.6.) and (2.7.) defining the scheme may be rearranged as follows:

$$q_\ell^{(k+1)} = q_\ell^{(k)} + e_\ell^{(k)} - e_{\ell-1}^{(k+1)}$$
$$e_\ell^{(k+1)} = e_\ell^{(k)} \frac{q_{\ell+1}^{(k)}}{q_\ell^{(k+1)}}$$

In this way quotients are formed only with the quantities $q_\ell^{(k)}$ which do not tend to zero and the qd-scheme is generated row by row. This is called the **progressive form** of the qd-algorithm. The problem of getting started is solved when we also calculate $q_\ell^{(k+1)}$ and $e_\ell^{(k+1)}$ for negative values of k, i.e. if we calculate the extended qd-scheme as described in section 3 of chapter II. In [29 pp.641-654] it is also shown how to deal with the case in which several poles have the same modulus.

To conclude this section we illustrate the preceeding theorem with the values $q_1^{(m)}$ and $e_1^{(m)}$ for

$$f(x) = \frac{\sin x}{0.1 - x}$$

which is an ultimately 1-normal function. Clearly with $w_0 = 0$, $w_1 = 0.1$ and $w_2 = \infty$,

$$\lim_{m \to \infty} q_1^{(m)} = 10 = \frac{1}{w_1}$$

and

$$\lim_{m \to \infty} e_1^{(m)} = 0$$

Table 4.18.

$q_1^{(1)}$	=	0.10000000000000E+02	$e_1^{(1)}$	=	-0.16666666666667E-01
$q_1^{(2)}$	=	0.99833333333333E+01	$e_1^{(2)}$	=	0.16666666666667E-01
$q_1^{(3)}$	=	0.10000000000000E+02	$e_1^{(3)}$	=	0.83472454091016E-05
$q_1^{(4)}$	=	0.10000008347245E+02	$e_1^{(4)}$	=	-0.83472454091016E-05
$q_1^{(5)}$	=	0.10000000000000E+02	$e_1^{(5)}$	=	-0.19874377699125E-08
$q_1^{(6)}$	=	0.99999999980126E+01	$e_1^{(6)}$	=	0.19874377699125E-08
$q_1^{(7)}$	=	0.10000000000000E+02	$e_1^{(7)}$	=	0.27600144392181E-12
$q_1^{(8)}$	=	0.10000000000000E+02	$e_1^{(8)}$	=	-0.27600144392181E-12
$q_1^{(9)}$	=	0.10000000000000E+02			

2.5. The generalized qd-algorithm.

The function $f(x)$ can also be given by its Newton series

$$f(x) = \sum_{i=0}^{\infty} f[x_0, \ldots, x_i] B_i(x)$$

where the $f[x_0, \ldots, x_i]$ are divided differences with possible coalescence of points and where

$$B_i(x) = \prod_{j=1}^{i} (x - x_{j-1})$$

If we define the generalized Hankel-determinants

$$\overline{H}_{m,n+1} = \begin{vmatrix} f[x_n, \ldots, x_m] & f[x_n, \ldots, x_{m+1}] & \cdots & f[x_n, \ldots, x_{n+m}] \\ f[x_{n-1}, \ldots, x_m] & \cdots & & f[x_{n-1}, \ldots, x_{n+m}] \\ \vdots & & & \vdots \\ f[x_0, \ldots, x_m] & \cdots & & f[x_0, \ldots, x_{n+m}] \end{vmatrix}$$

with $\overline{H}_{m,0} = 1$ then we call the Newton series **ultimately k-normal** if for every n with $0 \le n \le k$ there exists an integer M_n such that for $m \ge M_n$ the determinant $H_{m,n}$ is non-zero.

Theorem 4.7.

Let the sequence of interpolation points $\{x_0, x_1, x_2, \ldots\}$ be asymptotic to the sequence $\{z_0, z_1, \ldots, z_j, z_0, z_1, \ldots, z_j, z_0, z_1, \ldots, z_j, \ldots\}$ in the sense that

$$\lim_{k \to \infty} x_{k(j+1)+i} = z_i \qquad i = 0, \ldots, j$$

Let the function $f(x)$ be meromorphic in

$$B(z_0, \ldots, z_j, r) = \{x \in \mathbb{C} \mid |(x - z_0)(x - z_1) \ldots (x - z_j)| \le r\}$$

and analytic in the sequence of points $\{x_i\}_{i \in \mathbb{N}}$ and let the poles w_i of f in $B(z_0, \ldots, z_j, r)$ be numbered such that for

$$\omega_i = |(w_i - z_0) \ldots (w_i - z_j)|$$

we have

$$0 < \omega_1 \le \omega_2 \le \ldots < r$$

where the poles are counted with their multiplicities. If the Newton series is ultimately k-normal for some integer $k > 0$, then the generalized qd-scheme associated with this series has the following properties :

a) for each n such that $0 < n \leq k$ and such that $\omega_{n-1} < \omega_n < \omega_{n+1}$ where $\omega_0 = 0$ and, if f has exactly k poles, $\omega_{k+1} = r$, we have

$$\lim_{m \to \infty} q_n^{(m)} = \frac{1}{w_n - x_0}$$

b) for each n such that $0 < n \leq k$ and such that $\omega_n < \omega_{n+1}$, we have

$$\lim_{m \to \infty} e_n^{(m)} = 0$$

For the proof we refer to [9].

§3. Initial value problems.

Consider the following first order ordinary differential equation:

$$\frac{dy}{dx} = f(x,y) \quad \text{for} \quad x \in [a,b] \tag{4.10.}$$

with $y(a) = y_0$.

When we solve (4.10.) numerically, we do not look for an explicit formula giving $y(x)$ as a function of x but we content ourselves with the knowledge of $y(x_i)$ at several points x_i in $[a,b]$.

If we subdivide the interval $[a,b]$,

$$[a,b] = \bigcup_{i=1}^{k} [x_{i-1}, x_i]$$

where

$$x_i = a + ih \qquad i = 0, \ldots, k$$

with

$$h = \frac{b-a}{k} \quad \text{for} \quad k > 0$$

then we can calculate approximations y_{i+1} for $y(x_{i+1})$ by constructing local approximations for the solution $y(x)$ of (4.10.) at the point x_i. We restrict ourselves now to methods based on the use of nonlinear approximations.

3.1. The use of Padé approximants.

Let us try the following technique. If $s_i(x)$ is the Padé approximant of a certain order for $y(x)$ at x_i then we can put

$$y_{i+1} = s_i(x_{i+1})$$

which is an approximation for $y(x_{i+1})$. For the calculation of $s_i(x)$ we would need the Taylor series expansion of $y(x)$ at x_i, in other words

$$y(x) = y(x_i) + (x - x_i)f(x_i, y(x_i)) + \frac{(x - x_i)^2}{2} f'(x_i, y(x_i)) + \ldots$$

Since the exact value of $y(x_i)$ is not known itself, but only approximately by y_i, this Taylor series development is not known and hence this technique cannot be applied. However, we can proceed as follows. Consider the power series

$$y_i + (x - x_i)f(x_i, y_i) + \frac{(x - x_i)^2}{2} f'(x_i, y_i) + \ldots$$

Let $r_i(x)$ be the Padé approximant of order (m, n) for this power series. If we put $x = x_{i+1}$, in other words $x - x_i = h$, we obtain

$$y_{i+1} = r_i(x_{i+1}) \qquad i = 0, \ldots, k - 1$$

Hence we can write

$$y_{i+1} = y_i + hg(x_i, y_i, h) \qquad i = 0, \ldots, k - 1 \qquad (4.11.)$$

where g is determined by r_i.

Such a technique uses only the value of x_i and y_i to determine y_{i+1}. Consequently such methods are called **one-step** methods. Moreover (4.11.) is an **explicit** method for the calculation of y_{i+1}.

It is called a method of **order** p if the Taylor series expansion for $g(x, y, h)$ satisfies

$$y(x_{i+1}) - y(x_i) - hg(x_i, y(x_i), h) = 0(h^{p+1})$$

Clearly (4.11.) is a method of order $(m + n)$ if $r_i(x)$ is a normal Padé approximant. The convergence of (4.11.) follows if $g(x, y, h)$ satisfies the conditions of the following classical theorem [30 p. 71].

Theorem 4.8.

Let the function $g(x, y, h)$ be continuous and let there exist a constant L such that

$$|g(x, z, h) - g(x, y, h)| \leq L|z - y|$$

then the relation

$$g(x, y, 0) = f(x, y)$$

is a necessary and sufficient condition for the convergence of the method (4.11.), meaning that for fixed $x \in [a, b]$.

$$\lim_{\substack{h \to 0 \\ x = x_{i(h)}}} y_{i(h)} = y(x)$$

From the fact that r_i is a Padé approximant it follows that the relation $g(x, y, 0) = f(x, y)$ is always satisfied (see problem (8)).

The case $n = 0$ results in the classical Taylor series method for the problem (4.10.). If we take $m = n = 1$ we get

$$y_{i+1} = y_i + h\left[\frac{2f^2(x_i, y_i)}{2f(x_i, y_i) - hf'(x_i, y_i)}\right] \tag{4.12.}$$

If $y(x)$ is a rational function itself, then using (4.11.) we get the exact solution

$$y_{i+1} = y(x_{i+1})$$

at least theoretically, if the degrees of numerator and denominator of $r_i(x)$ are chosen in an appropriate way.

Techniques based on the use of Padé approximants can be interesting if we consider **stiff** differential equations, i.e. if $\frac{\partial f(x, y)}{\partial y}$ has a large negative real part [20]. An example of such a problem is the equation

$$\frac{dy}{dx} = \lambda y \tag{4.13.}$$

with $\text{Re}(\lambda)$ large and negative. Since the exact solution of (4.13.) is

$$y(x) = e^{\lambda x}$$

we have

$$\lim_{x \to \infty} y(x) = \lim_{x \to \infty} e^{\lambda x} = 0$$

and we want our approximations y_i to behave in the same way. Dahlquist [15] defined a method to be **A-stable** if it yields a numerical solution of (4.13.) with $\text{Re}(\lambda) < 0$ which tends to zero as $i \to \infty$ for any fixed positive h.

He also proved that there are no A-stable explicit linear one-step methods. Take for instance the method of Euler $(m = 1, n = 0)$:

$$y_{i+1} = y_i + hf(x_i, y_i)$$

We get

$$\begin{aligned} y_{i+1} &= (1 + h\lambda)y_i \\ &= (1 + h\lambda)^{i+1} y_0 \end{aligned}$$

Clearly

$$\lim_{i \to \infty} y_{i+1} = y_0 \lim_{i \to \infty} (1 + h\lambda)^{i+1} = 0$$

only if

$$|1 + h\lambda| < 1$$

So for large negative λ the steplength h has to be intolerably small before acceptable accuracy is obtained. In practice h is so small that round-off errors and computation time become critical. The problem is to develop methods that do not restrict the stepsize for stability reasons.

If (4.11.) results from the use of the Padé approximant of order (m, m), $(m, m + 1)$ or $(m, m + 2)$ then one gets an A-stable method [16]. This can be seen as follows. If $f(x, y) = \lambda y$ then $r_i(x)$ is the Padé approximant for the power series

$$y_i(1 + (x - x_i)\lambda + \frac{(x - x_i)^2}{2!}\lambda^2 + \ldots) = y_i \; e^{(x - x_i)\lambda}$$

Hence

$$y_{i+1} = r_{m,n}(h\lambda) \, y_i$$

with $h = x_{i+1} - x_i$, where $r_{m,n}(x)$ is the Padé approximant of order (m, n) for e^x. A-stability now follows from the following theorem.

Theorem 4.9.

If $m = n$ or $m = n - 1$ or $m = n - 2$ then the Padé approximant of order (m, n) for e^x satisfies

$$|r_{m,n}(x)| < 1 \quad \text{for} \quad \text{Re}(x) < 0$$

For the proof we refer to [3, 16].

3.2. The use of rational interpolants.

It is clear that if the interpolation conditions are spread over several points, then the computation of y_{i+1} will need several $x_{i-\ell}$ and $y_{i-\ell}(\ell = 0, 1, \ldots)$. Such methods are called **multistep** methods.

Let $r_i(x)$ be the rational Hermite interpolant of order (m, n) satisfying

$$r_i(x_{i-\ell}) = y_{i-\ell} \qquad \ell = 0, \ldots, j$$
$$r_i^{(t)}(x_{i-\ell}) = f^{(t-1)}(x_{i-\ell}, y_{i-\ell}) \qquad \ell = 0, \ldots, j \ \text{ and } \ t = 1, \ldots, s$$

where

$$(j+1)(s+1) = m+n+1$$

Here

$$f^{(t)}(x,y) = \frac{d^t f(x,y)}{dx} = \frac{\partial f^{(t-1)}(x,y)}{\partial x} + \frac{\partial f^{(t-1)}(x,y)}{\partial y} f(x,y)$$

Then an approximation for $y(x_{i+1})$ can be computed by putting

$$y_{i+1} = r_i(x_{i+1})$$

This is a nonlinear **explicit** multistep method. A two-step formula is obtained for instance by putting $j = 1$, $m = 2$, $n = 1$, and $s = 1$ and using theorem 3.17.:

$$y_{i+1} = \frac{4}{3}y_i - \frac{1}{3}y_{i-1} + \frac{2h}{9}(2f_i + f_{i-1}) + \frac{4h^2}{9} \frac{(f_i - f_{i-1})^2}{3(y_i - y_{i-1}) - h(f_i + 2f_{i-1})} \qquad (4.14.)$$

where $f_i = f(x_i, y_i)$ and $f_{i-1} = f(x_{i-1}, y_{i-1})$.
We can also derive **implicit** methods which require an approximation y_{i+1} for the calculation of y_{i+1} itself, by demanding

$$r_i(x_{i+1}) = y_{i+1}$$
$$r_i(x_{i-\ell}) = y_{i-\ell} \qquad \ell = 0, \ldots, j$$
$$r_i^{(t)}(x_{i-\ell}) = f^{(t-1)}(x_{i-\ell}, y_{i-\ell}) \qquad \ell = -1, 0, \ldots, j \ \text{ and } \ t = 1, \ldots, s$$

where

$$(j+2)(s+1) - 1 = m+n+1$$

For $m = 1 = n$, $j = 0$ and $s = 1$ we get the formula

$$y_{i+1} = y_i + h^2 \frac{f(x_{i+1}, y_{i+1})f(x_i, y_i)}{y_{i+1} - y_i} \qquad (4.15.)$$

For more information concerning such techniques we refer to [36]. Remark that multistep methods are never selfstarting. Both explicit and implicit $(j + 1)$- step methods are of the form

$$y_{i+1} = \sum_{\ell=0}^{j} \alpha_\ell y_{i-\ell} + hg(x_{i+1}, \ldots, x_{i-j}, y_{i+1}, \ldots, y_{i-j}, h)$$

and they have **order** p if the Taylor series expansion of g satisfies

$$y(x_{i+1}) - \sum_{\ell=0}^{j} \alpha_\ell y(x_{i-\ell}) - hg(x_{i+1}, \ldots, x_{i-j}, y(x_{i+1}), \ldots, y(x_{i-j}), h) = 0(h^{p+1})$$

Hence (4.14.) is a third-order method if the starting values are third-order and (4.15.) is second-order.

When applied to a stiff differential equation one should keep in mind that linear multistep methods are not A-stable if their order is greater than two.

The following result is helpful if $\frac{\partial f(x,y)}{\partial y}$ is real and negative. We know that we can write for problem (4.13.)

$$y(x) = y(x_{i-j})e^{(x-x_{i-j})\lambda}$$

with $\text{Re}(\lambda)$ large and negative. Hence it is interesting to take a closer look at rational Hermite interpolants for $\exp(x)$ in some real and negative interpolation points and also in 0.

Theorem 4.10.

If $r_i(x) = p_i(x)/q_i(x)$ with $\partial p_i \leq m$ and $\partial q_i \leq n$ is such that

a) $r_i^{(t)}(0) = 1 = \exp^{(t)}(0)$ for $0 \leq t \leq m + n - \ell$ with $\ell \leq m$

b) $r_i(\xi_\ell) = \exp(\xi_\ell)$ for $\xi_\ell < 0$, $1 \leq \ell \leq j$ with $\xi_\ell \neq \xi_k$, $1 \leq \ell \neq k \leq j$

then $|r_i(x)| < 1$ if $m \leq n$ and x is real and negative.

For the proof we refer to [32].

3.3. Predictor-corrector methods.

For a solution of the initial value problem (4.10.) we have

$$y'(x) = f(x, y(x)) \text{ for every } x \text{ in } [a, b]$$

If we integrate this equation on the interval $[x_{i-j}, x_{i+\ell}]$ with $j, \ell \geq 0$ we get

$$y(x_{i+\ell}) - y(x_{i-j}) = \int_{x_{i-j}}^{x_{i+\ell}} f(x, y(x))dx$$

Now f can be replaced by an interpolating function, through the points $(x_i, f(x_i, y_i))$, $(x_{i-1}, f(x_{i-1}, y_{i-1}))$, ..., which is easily integrated.
If $\ell = 1$ we get **predictor-methods** because they are explicit. If $\ell = 0$ we get **corrector-methods** because the value of y_i is needed for the computation of y_i. These implicit formulas can be used to update an estimate of $y(x_i)$ iteratively. When f is replaced by an interpolating polynomial we get the well-known methods of Adams-Bashforth ($j = 0$ and $\ell = 1$) and Adams-Moulton ($j = 1$ and $\ell = 0$). When f is replaced by an interpolating rational function we can get nonlinear formulas of predictor or corrector type.
Since the rational interpolant must be integrated, it is not recommendable to choose rational functions with a denominator of high degree.

3.4. Numerical results.

Let us compare two Taylor series methods ($m = 2$ and $m = 3$) with the explicit method (4.12.) for the solution of the equation

$$y' = 1 + y^2 \quad \text{for} \ x \geq 0$$
$$y(0) = 1$$

The theoretical solution is $y = tg(x + \pi/4)$. We take the steplength $h = 0.05$. As can be seen in table 4.19. the second order rational method gives even better results than the third order Taylor series method, a fact which can be explained by the singularity of the solution $y(x)$ at $x = \frac{\pi}{4}$.
To illustrate A-stability we will compare Euler's method ($m = 1, n = 0$) with the formulas (4.12.), (4.14.) and (4.15.) for the equation

$$y' = -25y \quad \text{for} \ x \in [0, 1]$$
$$y(0) = 1$$

The solution is known to be $y = \exp(-25x)$. For the results in table 4.20. we chose the steplength $h = 0.1$.
As expected Euler's solution blows up while formulas based on the use of a "diagonal" entry of the Padé table or the rational Hermite interpolation table for the exponential decay quite rapidly. Similar results would have been obtained if "superdiagonal" entries of the Padé table were used. Remark that formula

Table 4.19.

i	z_i	exact solution $\text{tg}(z_i + \frac{\pi}{4})$	Taylor series $m = 2$	Padé approximant $m = 1 = n$	Taylor series $m = 3$
1	0.05	0.110536D+01	0.110500D+01	0.110526D+01	0.110533D+01
2	0.10	0.122305D+01	0.122219D+01	0.122284D+01	0.122299D+01
3	0.15	0.135600D+01	0.135449D+01	0.135573D+01	0.135598D+01
4	0.20	0.150850D+01	0.150582D+01	0.150795D+01	0.150831D+01
5	0.25	0.168580D+01	0.168150D+01	0.168500D+01	0.168547D+01
6	0.30	0.189577D+01	0.188896D+01	0.189462D+01	0.189522D+01
7	0.35	0.214975D+01	0.213894D+01	0.214811D+01	0.214883D+01
8	0.40	0.246496D+01	0.244751D+01	0.246261D+01	0.246335D+01
9	0.45	0.286888D+01	0.283980D+01	0.286543D+01	0.286594D+01
10	0.50	0.340822D+01	0.335737D+01	0.340298D+01	0.340248D+01
11	0.55	0.416936D+01	0.407398D+01	0.416097D+01	0.415703D+01
12	0.60	0.533186D+01	0.513307D+01	0.531720D+01	0.530131D+01
13	0.65	0.734044D+01	0.685144D+01	0.731087D+01	0.724568D+01
14	0.70	0.116814D+02	0.100697D+02	0.116019D+02	0.112431D+02
15	0.75	0.282383D+02	0.177676D+02	0.277486D+02	0.232131D+02

Table 4.20.

i	x_i	exact solution $\exp(-25x_i)$	Euler	$m=1=n$ explicit one-step	$m=2, n=1$ explicit multistep	$m=1=n$ implicit one-step
1	0.1	0.820850D−01	−0.150000D+01	−0.111111D+00	0.820850D−01	0.123047D+00
2	0.2	0.673795D−02	0.225000D+01	0.123457D−01	0.840728D−01	0.151407D−01
3	0.3	0.553084D−03	−0.337500D+01	−0.137174D−02	−0.542641D−01	0.186302D−02
4	0.4	0.453999D−04	0.506250D+01	0.152418D−03	−0.494740D+00	0.229239D−03
5	0.5	0.372665D−05	−0.759375D+01	−0.169351D−04	−0.252173D+00	0.282073D−04
6	0.6	0.305902D−06	0.113906D+02	0.188168D−05	0.314955D+00	0.347083D−05
7	0.7	0.251100D−07	−0.170859D+02	−0.209075D−06	0.102175D+01	0.427077D−06
8	0.8	0.206115D−08	0.256289D+02	0.232306D−07	0.169137D+00	0.525507D−07
9	0.9	0.169190D−09	−0.384434D+02	−0.258117D−08	−0.191596D+00	0.646622D−08
10	1.0	0.138879D−10	0.576650D+02	0.288797D−09	−0.698115D+00	0.795651D−09

(4.14.) based on a "subdiagonal" rational approximation is surely not producing an A-stable method. To obtain the results of table 4.20. by means of (4.14.) a second starting value y_1 was necessary. We took $y_1 = \exp(-2.5) = y(x_1)$. For (4.15.) the expression $f(x_i, y_i) = -25y_i$ was substituted and y_{i+1} was solved from the quadratic equation.

All these schemes can be coupled to mesh refinement and the use of extrapolation methods. If an asymptotic error expansion of $y(x_i)$ in powers of h exists, then the convergence of the sequence of approximations for $y(x_i)$, with x_i fixed, obtained by letting the stepsize decrease, can be accelerated by the use of techniques described in section 1 [23].

3.5. Systems of first order ordinary differential equations.

Nonlinear techniques can also be used to solve a system of first order ordinary differential equations

$$\frac{dz_j}{dx} = f_j(x, z_1, z_2, \ldots, z_k) \qquad j = 1, \ldots, k$$
$$z_j(a) = z_{j,0}$$

where the values $z_{j,0}$ and the functions f_j are given for $j = 1, \ldots, k$. Several approaches are possible.

If we introduce vectors

$$Y(x) = \begin{pmatrix} z_1(x) \\ z_2(x) \\ \vdots \\ z_k(x) \end{pmatrix} \qquad F(x) = \begin{pmatrix} f_1(x, Y(x)) \\ f_2(x, Y(x)) \\ \vdots \\ f_k(x, Y(x)) \end{pmatrix}$$

then one method is to approximate the solution componentwise using similar techniques as in the preceding sections. So for instance (4.12.) becomes

$$Y(x_{i+1}) \simeq Y_{i+1} = Y_i + h \left[\frac{2F^2(x_i, Y_i)}{2F(x_i, Y_i) - hF'(x_i, Y_i)} \right] \qquad (4.16.)$$

where

$$F'(x, Y(x)) = \begin{pmatrix} f_1'(x, Y(x)) \\ f_2'(x, Y(x)) \\ \vdots \\ f_k'(x, Y(x)) \end{pmatrix}$$

and

$$Y_0 = \begin{pmatrix} z_{1,0} \\ z_{2,0} \\ \vdots \\ z_{k,0} \end{pmatrix}$$

and the addition and multiplication of vectors is performed componentwise. In other words (4.16.) is equivalent with

$$z_j(x_{i+1}) \simeq z_{j,i+1} =$$

$$z_{j,i} + h \frac{2 f_j^2(x_i, z_{1i}, \ldots, z_{ki})}{2 f_j(x_i, z_{1i}, \ldots, z_{ki}) - h f_j'(x_i, z_{1i}, \ldots, z_{ki})} \qquad j = 1, \ldots, k$$

For more information on such techniques we refer to [56, 35, 38].

Another approach is not based on componentwise approximation of the solution vector $Y(x)$ but is more vectorial in nature. Examples of such methods and a discussion of their properties is given in [57, 7, 27, 13].

The nonlinear techniques introduced here can also be used to solve higher order ordinary differential equations and boundary value problems because these can be rewritten as systems of first order ordinary differential equations. Again some of the nonlinear techniques prove to be especially useful if we are dealing with stiff problems. A system of differential equations

$$\frac{dY}{dx} = F(x, Y)$$
$$Y(a) = Y_0$$

with

$$\frac{dY}{dx} = \begin{pmatrix} \dfrac{dz_1}{dx} \\ \dfrac{dz_2}{dx} \\ \vdots \\ \dfrac{dz_k}{dx} \end{pmatrix}$$

is called **stiff** if the matrix

$$\begin{pmatrix} \dfrac{\partial f_1(x,Y)}{\partial z_1} & \cdots & \dfrac{\partial f_1(x,Y)}{\partial z_k} \\ \vdots & & \vdots \\ \dfrac{\partial f_k(x,Y)}{\partial z_1} & \cdots & \dfrac{\partial f_k(x,Y)}{\partial z_k} \end{pmatrix}$$

has eigenvalues with small and large negative real part.
Consider for example

$$\begin{cases} \dfrac{dz_1}{dx} = 998z_1(x) + 1998z_2(x) & z_1(0) = 1 \\[2mm] \dfrac{dz_2}{dx} = -999z_1(x) - 1999z_2(x) & z_2(0) = 0 \end{cases}$$

The solution is

$$z_1(x) = 2e^{-x} - e^{-1000x}$$
$$z_2(x) = -e^{-x} + e^{-1000x}$$

so that again

$$\lim_{x \to \infty} z_1(x) = 0 = \lim_{x \to \infty} z_2(x)$$

where both z_1 and z_2 contain fast and slow decaying components. For a discussion of stiff problems we refer to [20 pp.209-222].

§4. Numerical integration.

Consider $I = \int_a^b f(x)dx$. Many methods to calculate approximate values for I are based on replacing f by a function which can easily be integrated. The classical Newton-Cotes formulas are obtained in this way: f is replaced by an interpolating polynomial and hence I is approximated by a linear combination of function values.

In some cases the values of the derivatives of $f(x)$ are also taken into consideration and then linear combinations of the values of $f(x)$ and its derivatives at certain points are formed to approximate the value I of the integral. This is for instance the case if polynomial Hermite interpolation is used. In many cases the linear methods for approximating I give good results. There are however situations, for example if f has singularities, for which linear methods are unsatisfactory. So one could try to replace f by a rational function r and consider

$$\int_a^b r(x)dx$$

as an approximation for I. But rational functions are not that easy to integrate unless the poles of r are known and the partial fraction decomposition can be formed. Hence we use another technique.

Let us put

$$y(x) = \int_a^x f(t)dt$$

Then

$$I = y(b)$$

If f is Riemann integrable on $[a, b]$, then y is continuous on $[a, b]$. If f is continuous on $[a, b]$, then y is differentiable on $[a, b]$ with

$$y'(x) = f(x) \quad \text{and} \quad y(a) = 0$$

So I can be considered as the solution of an initial value problem and hence the techniques from the previous section can be used. We group them in different categories.

4.1. Methods using Padé approximants.

Let us partition the interval $[a, b]$ with steplength $h = (b - a)/k$ and write

$$x_i = a + ih \qquad i = 0, \ldots, k$$

and

$$t_{1,i}(h) = y_i + hf(x_i) + \frac{h^2}{2!}f'(x_i) + \frac{h^3}{3!}f''(x_i) + \ldots \qquad (4.17.)$$

where y_i approximates $y(x_i) = \int_a^{x_i} f(t)dt$.
If r_i is the Padé approximant of order (m, n) for $t_{1,i}(h)$ then we can put

$$y_{i+1} = r_i(h)$$

and consider y_k as an approximation for I. In this way

$$y_{i+1} = y_i + hg(x_i, h) \qquad i = 0, \ldots, k - 1 \qquad (4.18.)$$

which means

$$\int_{x_i}^{x_{i+1}} f(t)dt \simeq hg(x_i, h)$$

If $m = 1 = n$ we can easily read from (4.12.) that (4.18.) results in

$$\int_{x_i}^{x_{i+1}} f(t)dt \simeq h\frac{2f^2(x_i)}{2f(x_i) - hf'(x_i)} \qquad (4.19.)$$

Formulas like (4.18.) use derivatives of $f(x)$ and are nonlinear if $n > 0$. From the previous section we know that (4.18.) is exact, in other words that $y_k = I$, if $y(x)$ is a rational function with numerator of degree m and denominator of degree n. For $n = 0$ formula (4.18.) is exact if $y(x)$ is a polynomial of degree m, i.e. $f(x)$ is a polynomial of degree $m - 1$. The obtained integration rule is then said to be of **order** $m - 1$.
The convergence of formula (4.18.) is described in the following theorem which is only a reformulation of theorem 4.8.

Theorem 4.11.

Let y_i be defined by (4.18.). Then

$$\lim_{\substack{h \to 0 \\ x = x_{i(h)}}} y_{i(h)} = y(x) \quad \text{for fixed} \quad x \in [a, b]$$

if and only if $g(x, 0) = f(x)$.

Instead of (4.17.) one can also write

$$t_{1,i}(h) = y_i + h t_{2,i}(h)$$

with

$$t_{2,i}(h) = f(x_i) + \frac{h}{2!} f'(x_i) + \frac{h^2}{3!} f''(x_i) + \dots$$

and compute Padé approximants r_i of a certain order for $t_{2,i}(h)$. Let us take $m = 1 = n$. If we define

$$y_{i+1} = y_i + h r_i(h)$$

then we get

$$y_{i+1} = y_i + h^2 f(x_i) + h^2 \frac{3[f'(x_i)]^2}{6 f'(x_i) - 2h f''(x_i)}$$

4.2. Methods using rational interpolants.

If we again proceed as in the section on initial value problems we can construct nonlinear methods using information in more than one point. Since these methods are not self starting but need more than one starting value their use is somewhat limited and rather unpractical. An example of such a procedure is the following. Let $r_i(x)$ be the rational Hermite interpolant of order $(2, 1)$ satisfying

$$\begin{aligned} r_i(x_{i-j}) &= y_{i-j} \qquad j = 0, 1 \\ r_i'(x_{i-j}) &= f(x_{i-j}) \qquad j = 0, 1 \end{aligned}$$

and let

$$y_{i+1} = r_i(x_{i+1})$$

Then we know from formula (4.14.) that

$$y_{i+1} = \frac{4}{3}y_i - \frac{1}{3}y_{i-1} + \frac{2h}{9}(2f_i + f_{i-1}) + \frac{4h^2}{9}\frac{(f_i - f_{i-1})^2}{3(y_i - y_{i-1}) - h(f_i + 2f_{i-1})}$$

with $f_i = f(x_i)$ and $f_{i-1} = f(x_{i-1})$.

Often rational interpolants are preferred to Padé approximants for the solution of numerical problems because the use of derivatives of f is avoided. As mentioned, a drawback here is the necessity of more starting values. Another way to eliminate the use of derivatives, now without the need of more starting values, is the following.

4.3. Methods without the evaluation of derivatives.

One can replace derivatives of $f(x)$ in formula (4.18.) by linear combinations of function values of $f(x)$ without disturbing the order of the integration rule. To illustrate this procedure we consider the case $m = 1 = n$. Then [59]

$$g(x, h) = \frac{2f^2(x)}{2f(x) - hf'(x)}$$

We will compute constants α, β and γ such that for

$$t(x, h) = \frac{2f^2(x)}{\alpha f(x) + \beta f(x + \gamma h)}$$

we have

$$g(x, h) - t(x, h) = 0(h^{m+n}) = 0(h^2) \qquad (4.20.)$$

For

$$y_{i+1} = y_i + ht(x_i, h)$$

this would imply

$$y(x_{i+1}) - y_{i+1} = 0(h^{m+n+1}) = 0(h^3)$$

Condition (4.20.) is satisfied when

$$\alpha + \beta = 2$$
$$\beta\gamma = -1$$

In other words, for $\gamma \neq 0$,

$$\alpha = \frac{2\gamma + 1}{\gamma}$$

$$\beta = \frac{-1}{\gamma}$$

So

$$t(x, h) = \frac{2\gamma f^2(x)}{(2\gamma + 1)f(x) - f(x + \gamma h)}$$

For $\gamma = 1$ we get the integration rule

$$y_{i+1} = y_i + h\frac{2f^2(x_i)}{3f(x_i) - f(x_{i+1})}$$

In this way we approximate

$$\int_{x_i}^{x_{i+1}} f(t)dt \simeq h\frac{2f^2(x_i)}{3f(x_i) - f(x_{i+1})} \qquad (4.21.)$$

4.4. Numerical results for singular integrands.

We will now especially be interested in integrands regular in $[a, b]$ but with a singularity close to the interval of integration and on the other hand in integrands singular in a or b. The problem of integrating a function with several singularities within $[a, b]$ can always be reduced to the computation of a sum of integrals with endpoint singularities.

If $f(x)$ is singular in b then the value of the integral is defined by

$$I = \lim_{\epsilon \to 0} \int_a^{b-\epsilon} f(t)dt$$

and is assumed to exist.

We shall compare formula (4.19.) with Simpson's rule ($m = 2$, $n = 0$) and with a (2k)-point Gaussian quadrature rule that isolates the singularity in the weight function. If $f(t)$ can be written as $w(t)h(t)$ where $w(t)$ contains the endpoint singularity of $f(t)$ and $h(t)$ is regular then the approximation

$$I \approx w_1 h(t_1) + \ldots + w_{2k} h(t_{2k})$$

does not involve function evaluations in singular points. We use a (2k)-point formula because on $[x_i, x_{i+1}]$ for $i = 0, \ldots, k-1$ both (4.19.) and Simpson's rule

$$\int_{x_i}^{x_{i+1}} f(t)dt \approx \frac{h}{3}\left[f(x_i) + 4f\left(\frac{x_i + x_{i+1}}{2}\right) + f(x_{i+1})\right]$$

need two function evaluations.

Since f is singular in $b = x_k$, we take $f(x_k) = 0$ in Simpson's rule which means that the singularity is ignored. In (4.19.) the singularity of f is no problem since f and f' are only evaluated in $x_0, \ldots, x_{k-1}$ [58].

Our first numerical example is

$$\int_0^1 \frac{e^t}{(3 - e^t)^2} dt = 3.04964677\ldots$$

with a singularity in

$$t = \ln 3 = 1.09861228\ldots$$

and the second example is

$$\int_0^1 \frac{2(1 - t)\sin t + \cos t}{\sqrt{1 - t}} dt = 2$$

We shall also compare the different integration rules for the calculation of

$$\int_0^1 e^t dt = 1.71828182\ldots$$

which has a smooth integrand.

Because of the second example the weight function $w(t)$ in the Gaussian quadrature rule was taken to be

$$w(t) = \frac{1}{\sqrt{1 - t}}$$

All the computations were performed in double precision arithmetic and the double precision values for the weights w_i $(i = 1, \ldots, 2k)$ and the nodes t_i $(i = 1, \ldots, 2k)$ were taken from [1 pp. 916-919].

Remark that the nonlinear techniques behave better than the linear techniques in case of singular integrands. However, for smooth integrands such as in table 4.23., the classical linear methods give better results than the nonlinear techniques. Also, if the singularity can be isolated in the weight function such as in table 4.22., Gaussian quadrature rules are very accurate. In general, little accuracy is gained by using nonlinear techniques if other methods are available for the type of integrand considered [51].

As in the previous section all these schemes can be coupled to mesh refinement and extrapolation.

Table 4.21.

$$I = \int_0^1 \frac{e^t}{(3 - e^t)^2} dt = 3.04964677\ldots$$

	Gaussian quadrature	Simpson's rule	formula (4.19.)
$k = 4$	3.1734660	3.2806765	3.1006213
$k = 8$	3.0773081	3.0841993	3.0573798
$k = 16$	3.0564777	3.0531052	3.0510553

Table 4.22.

$$I = \int_0^1 \frac{2(1 - t)\sin t + \cos t}{\sqrt{1 - t}} dt = 2$$

	Gaussian quadrature	Simpson's rule	formula (4.19.)
$k = 4$	2.0000000	1.7504074	1.9033557
$k = 8$	2.0000000	1.8267581	1.8797832
$k = 16$	2.0000000	1.8786406	1.9048831

Table 4.23.

$$I = \int_0^1 e^t dt = 1.71828182\ldots$$

	Gaussian quadrature	Simpson's rule	formula (4.19.)
$k = 4$	1.7265746	1.7182842	1.7284991
$k = 8$	1.7204038	1.7182820	1.7206677
$k = 16$	1.7188196	1.7182818	1.7188592

§5. **Partial differential equations.**

Nonlinear techniques are not frequently used for the solution of partial differential equations. We will describe here a method based on the use of Padé approximants to solve the heat conduction equation which is a linear problem. For other illustrations we refer to [54, 48, 17, 22, 28]. All these techniques first discretize the problem so that the original partial differential equation is replaced by a system of equations which is nonlinear if the partial differential equation is. Another type of techniques which we do not consider here are methods which do not discretize the original problem but solve it iteratively by means of a procedure in which subsequent iteration steps are differentiable functions [13]. Linear techniques of this type are recommended for linear problems and nonlinear techniques can be used for nonlinear partial differential equations.

Let us now concentrate on the heat conduction equation.

Suppose we want to find a solution $u(x, t)$ of the linear problem

$$\frac{\partial u(x, t)}{\partial t} = \frac{\partial^2 u(x, t)}{\partial x^2} \qquad a < x < b, \ t > 0 \qquad (4.22.)$$

with

$$u(x, 0) = v(x)$$
$$u(a, t) = \alpha$$
$$u(b, t) = \beta$$

The domain $[a, b] \times [0, \infty)$ is replaced by a rectangular grid of points (x_i, t_j) with

$$x_i = a + ih \qquad i = 0, \ldots, k + 1$$
$$t_j = j\Delta t \qquad j = 0, 1, 2, \ldots$$

where

$$h = \frac{b - a}{k + 1}$$

and Δt is the discretization step for the time variable.

We first deal with the discretization in the space variable x and introduce

$$u_i(t) = u(x_i, t) \quad \text{for} \ t \geq 0$$

Then using central differences, (4.22.) can for instance be reduced to

$$\frac{du_i(t)}{dt} = \frac{u_{i+1}(t) - 2u_i(t) + u_{i-1}(t)}{h^2}$$
$$u_i(0) = v(x_i)$$

for $t > 0$ and $i = 1, \ldots, k$.
In general we can write

$$\begin{pmatrix} \dfrac{du_1(t)}{dt} \\ \vdots \\ \dfrac{du_k(t)}{dt} \end{pmatrix} = -A \begin{pmatrix} u_1(t) \\ \vdots \\ u_k(t) \end{pmatrix} \qquad (4.23.)$$

$$\begin{pmatrix} u_1(0) \\ \vdots \\ u_k(0) \end{pmatrix} = \begin{pmatrix} v(x_1) \\ \vdots \\ v(x_k) \end{pmatrix}$$

where A is a real symmetric positive definite $k \times k$ matrix and depends on the chosen approximation for the operator $\partial^2/\partial x^2$. If we introduce the notations

$$U(t) = \begin{pmatrix} u_1(t) \\ \vdots \\ u_k(t) \end{pmatrix} \qquad V = \begin{pmatrix} v(x_1) \\ \vdots \\ v(x_k) \end{pmatrix}$$

then the exact solution of (4.23.) is

$$U(t) = e^{-tA} V$$

where e^{-tA} is defined by

$$e^{-tA} = \sum_{\ell=0}^{\infty} \frac{(-t)^\ell}{\ell!} A^\ell$$

Using the discretization in the time variable t we can also write

$$U(t + \Delta t) = e^{-\Delta t \cdot A} U(t) \qquad (4.24.)$$

If $r_{m,n}(t) = p(t)/q(t)$ is the Padé approximant of order (m, n) for e^{-t} then (4.24.) can be approximated by

$$U(t + \Delta t) = [q(\Delta t.A)]^{-1} \ [p(\Delta t.A)] \ U(t) \qquad (4.25.)$$

Varga proved that for $n \geq m$ this is an unconditionally stable method [55], meaning that initial rounding errors remain within reasonable bounds as the computation proceeds independent of the stepsize used.
If $m = 1$ and $n = 0$ then (4.25.) means

$$U(t + \Delta t) = (I - \Delta t.A)U(t)$$

or equivalently if $A = (a_{i\ell})_{k \times k}$

$$u_i(t_{j+1}) = u_i(t_j) - \Delta t \sum_{\ell=1}^{k} a_{i\ell} u_\ell(t_j)$$

which is the well-known explicit method to solve (4.22.). The solution at level $t = t_{j+1}$ is determined from the solution at $t = t_j$.
For $m = 1 = n$ we obtain from (4.25.)

$$(I + \frac{1}{2}\Delta t.A)U(t + \Delta t) = (I - \frac{1}{2}\Delta t.A)U(t)$$

or equivalently

$$\frac{u_i(t_{j+1}) - u_i(t_j)}{\Delta t} = \frac{-1}{2}\left[\sum_{\ell=1}^{k} a_{i\ell}(u_\ell(t_{j+1}) + u_\ell(t_j))\right]$$

which is the method of Crank-Nicholson [50]. The operator $\partial^2/\partial x^2$ is replaced by the mean of an approximation for the partial second derivative at level $t = t_{j+1}$ and the same approximation at $t = t_j$.

§6. Integral equations.

As in the previous section we shall discuss linear equations for which the use of nonlinear methods is recommendable. Those interested in nonlinear integral equations are referred to [10, 12] where methods are indicated for their solution. If the integral equation is rewritten as a differential equation then techniques developed for the solution of initial value problems can also be used. We restrict ourselves to the discussion of an inhomogeneous Fredholm integral equation of the second kind (the unknown function f appears once outside the integral sign and once behind it):

$$f(x) - \lambda \int_a^b K(x,y)f(y)dy = g(x) \qquad x \in [a,b] \qquad (4.26.)$$

Here the kernel $K(x,y)$ and the inhomogeneous right hand side $g(x)$ are given real-valued continuous functions. Fredholm equations reduce to Volterra integral equations if the kernel $K(x,y)$ vanishes for $y > x$ which produces a variable integration limit.

6.1. Kernels of finite rank.

Formally the solution of (4.26.) can be written as a series. Put

$$f_0(x) = g(x)$$

and

$$f_{i+1}(x) = g(x) + \lambda \int_a^b K(x,y)f_i(y)dy \qquad i = 0,1,2,\ldots$$

If we define

$$K^1(x,y) = K(x,y)$$

and

$$K^i(x,y) = \int_a^b K(x,t)K^{i-1}(t,y)dt \qquad i = 2,3,\ldots$$

then (4.26.) reduces to

$$f(x) = \sum_{i=1}^{\infty} \lambda^i \int_a^b K^i(x,y)g(y)dy + g(x)$$

The series

$$g(x) + \sum_{i=1}^{\infty} \lambda^i \int_a^b K^i(x,y)g(y)dy \qquad (4.27.)$$

which is a power series in λ, is called the **Neumann series** of the equation (4.26.). Convergence of the Neumann series for certain values of λ depends on the properties of the kernel $K(x,y)$.

If $K(x,y)$ is bounded by

$$|K(x,y)| < M \quad \text{for} \quad (x,y) \in [a,b] \times [a,b]$$

then clearly the series (4.27.) converges uniformly to $f(x)$ in $[a,b]$ if

$$|\lambda| < \frac{1}{M(b-a)}$$

If Padé approximants in the variable λ are constructed for (4.27.) then they may have a larger convergence region than the series itself. Especially interesting is the case that the kernel is **degenerate**, in other words

$$K(x,y) = \sum_{i=1}^{k} X_i(x)Y_i(y)$$

with $\{X_i\}$ and $\{Y_i\}$ each linearly independent sets of functions. Such a kernel is also said to be of **finite rank** k. Let us try to determine $f(x)$ in this case. If we put

$$d_i = \int_a^b Y_i(y)f(y)dy$$

then (4.26.) can be written as

$$f(x) = g(x) + \lambda \sum_{i=1}^{k} d_i X_i(x) \qquad (4.28.)$$

Multiply (4.28.) by Y_j and integrate to get

$$\int_a^b f(y)Y_j(y)dy = \int_a^b g(y)Y_j(y)dy + \lambda \sum_{i=1}^{k} d_i \int_a^b X_i(y)Y_j(y)dy$$

or equivalently

$$d_j - \lambda \sum_{i=1}^{k} e_{ij} d_i = h_j \qquad (4.29.)$$

with

$$e_{ij} = \int_a^b X_i(y) Y_j(y) dy$$
$$\qquad\qquad i = 1, \ldots, k \ \text{ and } \ j = 1, \ldots, k$$
$$h_j = \int_a^b g(y) Y_j(y) dy$$

If we write

$$D(\lambda) = \begin{vmatrix} 1 - \lambda e_{11} & -\lambda e_{12} & \ldots & -\lambda e_{1k} \\ -\lambda e_{21} & 1 - \lambda e_{22} & \ldots & -\lambda e_{2k} \\ \vdots & & \ddots & \vdots \\ -\lambda e_{k1} & -\lambda e_{k2} & \ldots & 1 - \lambda e_{kk} \end{vmatrix}$$

which is the determinant of the coefficient matrix of system (4.29.), then $D(\lambda)$ is a polynomial in λ of degree at most k.

In case $D(\lambda) \neq 0$ the solution of (4.29.) is given by

$$d_i = \frac{\sum_{j=1}^{k} D_{ij} h_j}{D(\lambda)} \qquad i = 1, \ldots, k$$

with D_{ij} the minor of the $(i, j)^{th}$ element in $D(\lambda)$. So (4.28.) can be written as

$$f(x) = g(x) + \frac{\lambda}{D(\lambda)} \sum_{i=1}^{k} \left(\sum_{j=1}^{k} D_{ij} h_j \right) X_i(x) \qquad (4.30.)$$

which is a rational function in λ of degree at most k both in numerator and denominator. Since the series development of (4.30.) coincides with the Neumann series (4.27.) we know that the solution $f(x)$ is equal to the Padé approximant of order (k, k) for the Neumann series. So in this case the Padé approximant is the exact sum of the series because the sum is a rational function [2 p. 176].

6.2. Completely continuous kernels.

The equation (4.26.) can be rewritten as

$$(I - \lambda K)f = g \qquad (4.31.)$$

where the linear operators I and K are defined by

$$I : f(x) \rightarrow f(x) = If$$
$$K : f(x) \rightarrow \int_a^b K(x,y)f(y)dy = Kf$$

We shall consider square-integrable functions f with

$$\|f\| = \sqrt{\int_a^b |f(x)|^2 dx}$$

Suppose now that $\{f_i\}_{i \in \mathbb{N}}$ is a bounded converging sequence of functions. Then the operator K is said to be **completely continuous** if for all bounded $\{f_i\}_{i \in \mathbb{N}}$ the sequence $\{Kf_i\}_{i \in \mathbb{N}}$ contains a subsequence converging to some function $h(x)$ with

$$\|Kf_i - h\| = \sqrt{\int_a^b |Kf_i(x) - h(x)|^2 \, dx} \rightarrow 0 \quad \text{as} \quad i \rightarrow \infty$$

A basic property of completely continuous transformations is that they can be uniformly approximated by transformations of finite rank. Thus there is an infinite sequence $\{K_i\}_{i \geq 1}$ of kernels of finite rank i such that

$$\|(K - K_i)h\| < \epsilon_i \|h\| \quad \text{for all} \quad h \quad \text{with} \quad \|h\| < \infty$$

where

$$\lim_{i \to \infty} \epsilon_i = 0$$

When a completely continuous kernel K is replaced by K_i then the solution f_i of

$$(I - \lambda K_i)f_i = g$$

is given by the Padé approximant $r_{i,i}$ of order (i,i), as explained in the previous section. In case K is completely continuous the exact solution f of (4.31.) is a meromorphic function of λ [31 p. 31] and

$$\lim_{i \to \infty} f_i = f$$

in any compact set of the λ-plane except at the finite number of λ-poles of f [8]. Since

$$f_i = r_{i,i}$$

we consequently have

$$\lim_{i \to \infty} r_{i,i} = f$$

in any compact set of the λ-plane except at the finite number of λ-poles of f. Thus the solution f is the limit of a sequence of diagonal Padé approximants. This result is not very useful since each approximant in the sequence is derived from a different Neumann series, with kernel K_i and not with kernel K. However a similar result exists for the sequence $r_{i,i}$ derived from the Neumann series

$$g + \sum_{i=1}^{\infty} (\lambda K)^i g \qquad (4.32.)$$

where the operator K^i is defined by

$$K^i : g(x) \to \int_a^b K^i(x, y)g(y)dy$$

Theorem 4.12.

If $r_{i,i}$ is the Padé approximant of order (i, i) to the Neumann series (4.32.) of the integral equation (4.26.) with completely continuous kernel K then the solution f is given by

$$\lim_{i \to \infty} r_{i,i} = f$$

in any compact set of the λ-plane except at the finite number of poles of f and at limit points of poles of $r_{i,i}$ as $i \to \infty$.

More information on this subject can be found in [2 pp. 178-182].

Problems.

(1)   Show that with $\epsilon_0^{(n)} = a_n$ for $n \geq 0$,

$$\epsilon_2^{(n)} = \frac{a_{n+2}\, a_n - a_{n+1}^2}{a_{n+2} - 2a_{n+1} + a_n}$$
$$= a_{n+1} - \frac{\Delta a_n \Delta a_{n+1}}{\Delta^2 a_n}$$

which coincides with Aitken's Δ^2-process to accelerate the convergence of a sequence.

(2)   Give an algorithm similar to the one in section 1.4. for the calculation of $t_{i,i}$, but now based on the use of inverse differences instead of reciprocal differences.

(3)   Show that if the algorithm of Bulirsch and Stoer is used with the interpolation points x_i ($\lim_{i \to \infty} x_i = 0$) and the ρ-algorithm is used with the interpolation points $x_i' = 1/x_i$ ($\lim_{i \to \infty} x_i' = \infty$), then $t_{ii} = s_i^{(0)}$.

(4)   Compare the amount of additions and multiplications performed in the ϵ-algorithm and the qd-algorithm when used for convergence acceleration.

(5)   Derive the formulas (4.3.) and (4.4.) using inverse interpolation instead of direct interpolation.

(6)   Derive the formulas (4.8.) and (4.9.) based on the use of rational interpolants.

(7)   Organize the computation of d_{m+1} in section 2.3. for successive values of i such that a mininum number of operations is involved.

(8)   Prove that one-step explicit methods for the solution of initial value problems based on the use of Padé approximants are convergent if $g(x, y, h)$ given by (4.11.) is continuous and satisfies

$$|g(x, z, h) - g(x, y, h)| \leq L|z - y|$$

(9)   Check the formulas (4.12.) and (4.14.).

(10)   Write down formula (4.16.) for the solution of

$$\begin{cases} \dfrac{dz_1}{dx} = 998z_1(x) + 1998z_2(x) & z_1(0) = 1 \\[2mm] \dfrac{dz_2}{dx} = -999z_1(x) - 1999z_2(x) & z_2(0) = 0 \end{cases}$$

(11)   Construct a nonlinear numerical integration rule based on the use of Padé approximants of order $(2, 1)$. Afterwards eliminate the use of derivatives as explained in section 4.3.

Remarks.

(1)   Nonlinear methods can also be used for the solution of other numerical problems. We refer for instance to [40] where the solution of linear systems of equations is treated, to [19] for analytic continuation, to [34] for numerical differentiation and to [41] for the inversion of Laplace transforms.

(2)   An important link with the theory of numerical linear algebra is through QR-factorization. Rutishauser [47] proved that the determination of the eigenvalues of a square matrix A can be reduced to the determination of the poles of a rational function f built from the given matrix. In this way decomposition techniques for A to compute its eigenvalues are related to the qd-algorithm when used to compute poles of meromorphic functions.

(3)   Univariate and multivariate continued fractions and rational functions are also often used to approximate given functions. For univariate examples we refer to [33, 42]. The bivariate Beta function is a popular multivariate example because it has numerous singularities in a quite regular pattern. For numerical results we refer to [39, 26, 12].

(4)   As a conclusion we can say that every linear method has its nonlinear analogue. In case linear methods are inaccurate or divergent, it is recommendable to use a similar nonlinear technique. The price we have to pay for the ability of the nonlinear method to cope with the singularities is the programming difficulty to avoid division by small numbers.

References.

[1] *Abramowitz M.* and *Stegun I.*　Handbook of Mathematical functions. Dover publications, New York, 1968.

[2] *Baker G.* and *Gammel J.*　The Padé approximant in theoretical physics. Academic Press, New York, 1970.

[3] *Baker G.* and *Graves-Morris P.*　Padé approximants: basic theory. Encyclopedia of Mathematics and its applications: vol 13. Addison-Wesley, Reading, 1981.

[4] *Brezinski C.*　Algorithmes d'accélération de la convergence. Editions Technip, Paris, 1978.

[5] *Brezinski C.*　Application de l'ε-algorithme à la résolution des systèmes non linéaires. C. R. Acad. Sci. Paris Sér. A 271, 1970, 1174-1177.

[6] *Bulirsch R.* and *Stoer J.*　Fehlerabschätzungen und Extrapolation mit rationalen Funktionen bei Verfahren vom Richardson-Typus. Numer. Math. 6, 1964, 413-427.

[7] *Calvo M.* and *Mar Quemada M.*　On the stability of rational Runge Kutta methods. J. Comput. Appl. Math. 8, 1982, 289-292.

[8] *Chisholm J.*　Solution of linear integral equations using Padé Approximants. J. Math. Phys. 4, 1963, 1506-1510.

[9] *Claessens G.*　Convergence of multipoint Padé approximants. Report 77-26, University of Antwerp, 1977.

[10] *Clarysse T.*　Rational predictor-corrector methods for nonlinear Volterra integral equations of the second kind. In [60], 278-294.

[11] *Cuyt A.*　Accelerating the convergence of a table with multiple entry. Numer. Math. 41, 1983, 281-286.

[12] *Cuyt A.*　Padé approximants for operators: theory and applications. Lecture Notes in Mathematics 1065, Springer Verlag, Berlin, 1984.

[13] *Cuyt A.*　Padé approximants in operator theory for the solution of nonlinear differential and integral equations. Comput. Math. Appl. 6, 1982, 445-466.

[14] *Cuyt A.* and *Van der Cruyssen P.* Abstract Padé approximants for the solution of a system of nonlinear equations. Comput. Math. Appl. 9, 1983, 617-624.

[15] *Dahlquist G.* A special stability problem for linear multistep methods. BIT 3, 1963, 27-43.

[16] *Ehle B.* A-stable methods and Padé approximations to the exponential. SIAM J. Math. Anal. 4, 1973, 671-680.

[17] *Fairweather G.* A note on the efficient implementation of certain Padé methods for linear parabolic problems. BIT 18, 1978, 106-108.

[18] *Frame J.* The solution of equations by continued fractions. Amer. Math. Monthly 60, 1953, 293-305.

[19] *Gammel J.* Continuation of functions beyond natural boundaries. Rocky Mountain J. Math. 4, 1974, 203-206.

[20] *Gear C.* Numerical initial value problems in ordinary differential equations. Prentice-Hall Inc., New Yersey, 1971.

[21] *Genz A.* The approximate calculation of multidimensional integrals using extrapolation methods. Ph. D. in Appl. Math., University of Kent, 1975.

[22] *Gerber P.* and *Miranker W.* Nonlinear difference schemes for linear partial differential equations. Computing 11, 1973, 197-212.

[23] *Gragg W.* On extrapolation algorithms for ordinary initial value problems. SIAM J. Numer. Anal. 2, 1965, 384-403.

[24] *Gragg W.* and *Stewart G.* A stable variant of the secant method for solving nonlinear equations. SIAM J. Numer. Anal. 13, 1976, 889-903.

[25] *Graves-Morris P.* Padé approximants and their applications. Academic Press, London, 1973.

[26] *Graves-Morris P. , Hughes Jones R.* and *Makinson G.* The calculation of some rational approximants in two variables. J. Inst. Math. Appl. 13, 1974, 311-320.

[27] *Hairer E.* Nonlinear stability of RAT, an explicit rational Runge-Kutta method. BIT 19, 1979, 540-542.

[28] *Hall C.* and *Porsching T.*   Padé approximants, fractional step methods and Navier-Stokes discretizations. SIAM J. Numer. Anal. 17, 1980, 840-851.

[29] *Henrici P.*   Applied and computational complex analysis: vol. 1. John Wiley, New York, 1974.

[30] *Henrici P.*   Discrete variable methods in ordinary differential equations. John Wiley, New York, 1962.

[31] *Hoheisel G.*   Integral equations. Ungar, New York, 1968.

[32] *Iserles A.*   On the generalized Padé approximations to the exponential function. SIAM J. Math. Anal. 16, 1979, 631-636.

[33] *Kogbetliantz E.*   Generation of elementary functions. In [46], 7-35.

[34] *Kopal Z.*   Operational methods in numerical analysis based on rational approximation. In [37], 25-43.

[35] *Lambert J.*   Computational methods in ordinary differential equations. John Wiley, London, 1973.

[36] *Lambert J.* and *Shaw B.*   A method for the numerical solution of $y' = f(x, y)$ based on a self-adjusting non-polynomial interpolant. Math. Comp. 20, 1966, 11-20.

[37] *Langer R.*   On numerical approximation. University of Wisconsin Press, Madison, 1959.

[38] *Lee D.* and *Preiser S.*   A class of nonlinear multistep A-stable numerical methods for solving stiff differential equations. Internat. J. Comput. Math. 4, 1978, 43-52.

[39] *Levin D.*   On accelerating the convergence of infinite double series and integrals. Math. Comp. 35, 1980, 1331-1345.

[40] *Lindskog G.*   The continued fraction methods for the solution of systems of linear equations. BIT 22, 1982, 519-527.

[41] *Longman I.*   Use of Padé table for approximate Laplace transform inversion. In [25], 131-134.

[42] *Luke Y.*   The special functions and their approximations. Academic Press, New York, 1969.

[43] *Merz G.*   Padésche Näherungsbrüche und Iterationsverfahren höherer Ordnung. Computing 3, 1968, 165-183.

[44] *Nourein M.*   Root determination by use of Padé approximants. BIT 16, 1976, 291- 297.

[45] *Ortega J.* and *Rheinboldt W.*   Iterative solution of nonlinear equations in several variables. Academic Press, New York, 1970.

[46] *Ralston A.* and *Wilf S.*   Mathematical methods for digital computers. John Wiley, New York, 1960.

[47] *Rutishauser H.*   Der Quotienten-Differenzen Algorithmus. Z. Angew. Math. Phys. 5, 1954, 233-251.

[48] *Siemieniuch J.* and *Gladwell I.*   On time-discretizations for linear time-dependent partial differential equations. University of Manchester, Numer. Anal. Report 5, 1974.

[49] *Smith D.* and *Ford W.*   Numerical comparison of nonlinear convergence accelerators. Math. Comp. 38, 1982, 481-499.

[50] *Smith G.*   Numerical solution of partial differential equations. Oxford University Press, London, 1975.

[51] *Squire W.*   In defense of linear quadrature rules. Comput. Math. Appl. 7, 1981, 147-149.

[52] *Tornheim L.*   Convergence of multipoint iterative methods. J. Assoc. Comput. Mach. 11, 1964, 210-220.

[53] *Traub J.*   Iterative methods for the solution of equations. Prentice-Hall Inc., New York, 1964.

[54] *Varga R.*   Matrix iterative analysis. Prentice-Hall Inc., Englewood-Cliffs, 1962.

[55] *Varga R.*   On high order stable implicit methods for solving parabolic partial differential equations. J. Math. Phys. 40, 1961, 220-231.

[56] *Wambecq A.*   Nonlinear methods in solving ordinary differential equations. J. Comput. Appl. Math. 1, 1976, 27-33.

[57] *Wambecq A.*   Rational Runge-Kutta methods for solving systems of ordinary differential equations. Computing 20, 1978, 333-342.

[58] *Werner H.* and *Wuytack L.*   Nonlinear quadrature rules in the presence of a singularity. Comput. Math. Appl. 4, 1978, 237-245.

[59] *Wuytack L.*   Numerical integration by using nonlinear techniques. J. Comput. Appl. Math. 1, 1975, 267-272.

[60] *Wuytack L.*   Padé approximation and its applications. Lecture Notes in Mathematics 765, Springer, Berlin, 1979.

[61] *Wynn P.*   Acceleration techniques for iterated vector and matrix problems. Math. Comp. 16, 1962, 301-322.

[62] *Wynn P.*   Singular rules for certain non-linear algorithms. BIT 3, 1963, 175-195.

SUBJECT INDEX